Recording Tips for Engineers

For cleaner, brighter tracks

Second edition

Tim Crich

ELSEVIER

AMSTERDAM • BOSTON • HEIDELBERG • LONDON • NEW YORK • OXFORD
PARIS • SAN DIEGO • SAN FRANCISCO • SINGAPORE • SYDNEY • TOKYO
Focal Press is an imprint of Elsevier

Focal
Press

Focal Press is an imprint of Elsevier
Linacre House, Jordan Hill, Oxford OX2 8DP, UK
30 Corporate Drive, Suite 400, Burlington, MA 01803, USA

First edition published by Black Ink Publishing 2002
Second edition published by Elsevier 2005
Reprinted 2006

Notice
No responsibility is assumed by the publisher for any injury and/or damage to persons
or property as a matter of products liability, negligence or otherwise, or from any use
or operation of any methods, products, instructions or ideas contained in the material
herein. Because of rapid advances in the medical sciences, in particular, independent
verification of diagnoses and drug dosages should be made

British Library Cataloguing in Publication Data
A catalogue record for this book is available from the British Library

Library of Congress Cataloging-in-Publication Data
A catalog record for this book is available from the Library of Congress

ISBN–13: 978-0-240-51974-6
ISBN–10: 0-240-51974-4

For information on all Focal Press publications
visit our website at www.focalpress.com

Printed and bound in *The Netherlands*

06 07 08 09 10 10 9 8 7 6 5 4 3 2

Working together to grow
libraries in developing countries
www.elsevier.com | www.bookaid.org | www.sabre.org

ELSEVIER   BOOK AID International   Sabre Foundation

Contents

Contents

Chapter Three: The Drums Setup   41

Chapter Four: The Electric Guitar Setup   64

Contents

Chapter Nine: The Signal Routing   150

Chapter Ten: The Recording   159

Chapter Eleven: The Mixing   180

Contents

Chapter Twelve: The Digital Appendix 217

Index 231

List of illustrations

Acknowledgements

Ron 'Obvious' Vermeulen, Rick Eden, Bob Clearmountain, Bryan Adams, Bob Rock, Bruce Bartlett, Mike Collins, Beth Howard, Simon Andrews, Jeb Stuart-Bullock, Frank Filipetti, Bob Schwall, Alan Friedman, Mark Hermann, Noah Baron, Josiah Gluck, Eric LeMay, Paul McGrath, Blake Havard, Hal Armour, Rob Pyne, Tim King, Joan and Glenn Crich, Buck Crich, Matthew Crich, Cam and Inez McLean.

Special thanks to Ron, Bryan and Warehouse Studios. Extra special triple thanks to my wife Grace.

How to be a recording engineer

Many people choose to become recording engineers because they want to party all day. That's great for the first few decades, but soon they will realize that it is a complex job and an art. Understanding what all the knobs do is just the beginning. Being a recording engineer is not just about using electronic equipment, it is also about effectively capturing music, sound effects, or the spoken word. The studio is just a tool. Often so is the studio manager, but that's another story.

I invite you to go through this book, choose the things you find applicable, and leave the rest. There are a few items covered that are also in *Assistant Engineers Handbook* but overlap is kept to a minimum. Note that the term 'he' is used strictly for ease. There is ample work for everyone, he or she.

Some of the enclosed tips are from other engineers, some are standard studio practice, and some were learned while on the job. All of them have helped me in the studio, and I hope they will help you too.

Thank you very much for reading this book, and good luck as a recording engineer. It can be lucrative and it can be draining. Take the good with the bad, experiment, stay up late, and have fun.

May your shirts always be louder than your speakers.

Good luck.

Tim Crich

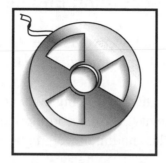

Chapter One

THE RECORDING ENGINEER

This book is for the recording engineer who knows what a compressor or limiter or equalizer does. He wants to know the tried and true methods of using all of the equipment presently available.

The recording engineer is responsible for keeping the session running smoothly, including setting up the control room, choosing the microphones, organizing the signal flow, choosing the track layout, getting the sounds and pressing the record button. Good sounds or bad, the buck stops with the recording engineer.

Becoming a Better Recording Engineer

• **Praise the lowered.** Work at lower volume levels. The sounds will be more accurate and ear fatigue will be minimized. If the level must be loud, get your sounds, insert your earplugs and turn it up. Occasionally listen at lower levels without the earplugs. There is nothing in the recording studio as important as your hearing. Longevity in the recording industry means good hearing for decades to come. Plus the loud level might wake up the producer.

• **Get musical.** Recording music is so much easier if you understand music. Music plays a key role in a vast majority of recordings so most clients prefer 'musical' engineers. If you don't play an instrument, buy a guitar or keyboard and learn some basic songs. While learning to play an instrument may seem daunting, you don't need to become a virtuoso player, you just need to grasp song structures and musical progressions. If you get musical, you get work.

• **Be consistent.** Quality is no accident. Success comes from working every day at your craft, and that requires hard work and dedication. You become what you practice. The ultimate goal is to be the recording engineer that everyone wants to use because of your ears, your expertise, your vibe, and your impressive collection of Ramones t-shirts.

• **I'm maintaining.** Keep your body well maintained or long hours will take their toll. Just like your car, if you give it the best gasoline you will get the best results. Eat healthy and drink enough water.

• **Be professional.** This is your craft, and you must work at it. I have seen engineers lose gigs because they got wasted and became an idiot. Do what I do – wait until your day off to start drinking at 7 a.m.

• **Don't get mad, get even.** An even temperament goes a long way. Mistakes and frustrations happen in all jobs, and in the long run, so what? A good engineer keeps the session at ease, especially during stressful times. Do you want clients and co-workers to remember you as the engineer who blows up, or the engineer who is a pro and can work around anything?

• **A breath of fresh air.** When you sit at the console next to someone for hours on end, a toothbrush, mouthwash and breath mints may be in order.

• **Fess up.** As the engineer, you are responsible for the content of the recording. If you make a mistake or erase something, say so. You will get more respect in the long run.

• **Discrete recording.** Discretion for an engineer means knowing when to crank the volume for a playback, when to be quiet and twiddle the knobs, and when to move on. As the engineer, you lead the session. The producer has the road map, but you drive the car.

• **Make it look good.** Some engineers go through their careers simply putting up a microphone and pressing the record button. Engineering is an art. Much like cooking and sex, presentation is part of the package.

• **So shut up already.** There's no reason to broadcast to everyone that you are manipulating an instrument or vocal sound. Just quietly do it, and if they ask, tell them any changes are minor. Announcing, 'I really had to use a lot of equalization on your vocal' helps no one. Just get the sound and, as Joe Perry said, 'Let the music do the talking.'

• **Record what the song requires.** If the song requires bagpipes, don't use something that sounds sort of like bagpipes, get the bagpipes. Whenever you compromise, you might save a couple of hours and a couple of dollars, but a mediocre substitute haunts you long after the money and time are forgotten. Money comes and goes but a recording is timeless. Especially the bagpipes.

• **Record an instrument how it's supposed to sound.** This may be obvious, but if you are recording an unfamiliar instrument, go into the studio and listen to the instrument being played. Maybe discuss how it's supposed to sound with the player. Some instruments are heavy within a certain frequency range and are frequency dependent. If the sound dwells within a limited frequency range, don't 'fix' it with processing – record it how it's supposed to sound. Then, if the natural sound of the instrument isn't working, maybe it's the wrong part or the wrong instrument, or both.

• **Get a good sound fast.** People lose perspective when the engineer takes three hours getting a bass sound. Unless it sounds horrible, move on.

• **Spend the most time on the most important factor of the record.** If the main part of the session is scheduled for vocals, don't spend hours on the drums.

• **A/B and see.** Once you process a sound, press the bypass button to compare that the processed sound is an improvement over the unprocessed sound.

• **Leave the solo button alone.** An instrument in solo sounds a lot different when the rest of the tracks are in the monitor mix.

• **Use microphone choice, setting, and placement over processing.** If you can get a better sound by slightly moving the microphone, do that before adding equalization and compression.

• **Commit to the sound.** The confident engineer says, 'This is the sound we want, let's record it,' rather than, 'Hey, how's this?' Unless someone really doesn't like the sound, everyone should go along. But you must be correct. Without confidence in the engineer, things can quickly deteriorate from 'Sounds great' to 'Gee I don't know, what do *you* think?'

• **Rule of thumb.** Make the guy who signs your check sound best.

Dealing with Clients

• **Respect the client.** Be punctual. By showing up late, you are saying 'Your project is not that interesting to me.' Clients want to be sure you take them seriously. If a few minor things go wrong, one guy always says, 'Yeah, and he's never on time either.'

As well, leave your watch at home, remember everybody's name, and consider every band or project you work on as 'the next big thing.'

• **Respect the music.** Keep your own funny versions of a client's songs to yourself. Be a professional and respect the people who wrote the lyrics. Don't suggest lyric changes, and never say 'Y'know, this song sounds just like...' That serves no creative purpose, and makes the songwriter look like a hack.

• **If you want loyalty in the music business, buy a dog.** Don't get too attached to a project. They will say they love you, love your engineering, are definitely going to use you next time, you're in the club, the sounds are brilliant. Next week you hear they are using another engineer. Well don't let it bug you. Do your job, take pride in it, and at the end of the day, realize that no matter what they promise, you don't have the gig until you're in the chair.

• **Save a copy of everything you do.** Who knows who the next major stars will be. As well, you can track your progress as an engineer over the years. Keep a copy – even if you think you will never use it again.

• **Seven days in the studio makes one weak.** Some clients will expect you to work long hours without a day off. This benefits no one. The eighteenth hour of the tenth day in a row is when mistakes happen. You want your clients to remember you for your skills as an engineer, not for erasing the kick drum due to fatigue. And once you start working long hours, the client expects it.

• **Although it may not be your job, keep in mind the budget of the project.** Spending tons of time on a part that really isn't that important might take away from precious mixing time. You can't explain away an average mix.

• **Watch it or I'll flatten your EQ.** If you don't get along with someone in the sessions, deal with it. Probably you (the lowly engineer) will go before he (the high and mighty musician) does. But, life isn't long enough to take abuse from anyone. But, if you're being ill-treated, give them a serious staring at, then walk.

• **Be the heavy.** Sometimes the engineer must also be the heavy, doing the unpleasant tasks when sessions get out of hand. State firmly and professionally, 'You can't smoke in the control room.' 'Please don't set your drink on the console.' 'You girls put your clothes on this instant!'

5

Getting Paid

• **Don't record for free.** Sometimes to get experience, junior engineers record bands for free. Don't. Engineering a project for 'nothing' makes the client think you are worth nothing. Even if it's a token payment of ten dollars for lunch, take it.

• **Due time or do time.** Independent engineers often get payment in full without any tax deductions. Because you are a professional, keep all receipts, notes of sessions, who paid what, and all work orders. You know the government and taxes. Man, they make a federal case out of everything.

• **A taxman attacks man.** Recording engineers have certain tax write-offs. Take advantage of these write-offs. Subscribe to all the applicable magazines. Buy lots of books, meals, tapes, guitar picks, earplugs, hairplugs, whatever. Write them all off.

• **Paper covers rock.** Professional-looking paperwork tells your clients you are a professional recording engineer, not some hack with a handwritten invoice and no business card. If you do business like a professional you will be treated like one.

• **Now be honest.** No matter how hard you work or how great your sound, always be honest and professional in your approach. The industry is small, and bad news travels fast.

Keeping Up

• **Install a DAW (Digital Audio Workstation) editing program on your computer.** Improve your ability to edit, equalize, compress and record. Before buying any studio software, go on-line and read some reviews, check out the website, and download demos. All product sellers have websites with demos available. Once armed with the facts, purchase and install the appropriate software and hardware.

Purchase a quality sound card. The cheap sound cards use low quality electronics that overload sooner.

• **Buy at least one good microphone.** You'll soon learn its characteristics and be able to compare it with other microphones. When you hear a different microphone next to it, you can say, 'That's a little brighter than mine, or the low end is cloudier.' Today's microphones are inexpensive and high quality.

Beware. The audio industry sometimes goes through 'fads' where every studio buys the newest thing, then realizes it isn't so hot. Buy equipment that will last for the long haul – and buy from a reputable dealer.

• **The joy of specs.** Read all the industry magazines to keep up with the latest technologies. Read all the studio manuals, attend trade shows, surf all the websites, listen to the sound industry CD ROMs, and check the available equalization and tone reference discs. Grasping all the workings in a modern recording studio can only help you. You can bet the big name recording engineers understand this stuff.

• **Use equipment to its max.** Don't be too nervous to try new things out. But unless the project has an unlimited budget, most clients don't want to spend hours on end waiting for you to try a groovy new idea. Come in on your own time to work on sounds and ideas, then have them in your arsenal. Lay one of these ideas on the client during the session. Respect soon follows.

Getting Work

• **The music business is tough.** Work is elusive and you have to hunt it down. Check out all the local studios. Leave a card. Try to get a rapport with certain studios. If you bring in a few bands, you may get a break on the cost of the studio. As well, if they know you and are familiar with your work, they may call you when they need an engineer.

• **Check your hearing.** Before you seriously become an active, working recording engineer, get your head, er...hearing examined. If your hearing is questionable, it ain't getting any better. It may be disconcerting if the client sees you adjusting your hearing aid in the session.

• **Have ears will travel.** Place an ad in the local music paper that you are available to record bands at a very reasonable rate. Go to clubs and talk to bands about recording. Print up a demo disc of some of your best work – even if you must book studio time to do it – and mail it out or hand it out to whomever may be interested.

Include a business card with a contact number. Don't scribble, 'This whole disc was recorded in half an hour in Dave's basement with no overdubs and lots of beer.' Use professional graphics.

• **I love the mall, I love them all.** Get to know as many people in the local scene as possible by hanging around the music and recording gear stores, going to shows, and supporting local artists. Small-time recording engineers, managers and local musicians become big name producers, studio owners and rock stars.

• **He shoots, he scores.** Do you play hockey, baseball, bowling, curling, tongue wrestling? Many cities have music industry sports teams. This is a good way to network in the recording industry. There is nothing like getting a little sympathy work, so maybe a puck in the head now and again will help your studio career.

• **Intensities in ten cities.** Attend the major audio shows and conventions such as the AES or NAMM. Go to these shows to learn what is on the horizon and to hang out with audio industry leaders.

• **Get outta town.** You may want to move to a locale that has lots of studios, like LA, NY, or Nashville. There are many secondary markets other than these three, but of course these ones are the main places. Note that even though there are more studios, there is more competition, and big cities aren't for everyone.

• **And on this team.** Just working as a recording engineer today might not bring in enough work. The competition can be fierce. So many engineers are teaming up with people such as a producers or mixers and starting their own production company. With the low cost of equipment, this may be a viable option for some.

• **Use your computer to its fullest capacity.** Use the Internet to access data on recording studios, new equipment, and newest techniques. There are many websites available to research available recording studios in your area, as well as any new techniques.

Keep a file on all the studios including a short list of the attributes of each studio, including the names of the studio manager and the studio owner.

Create a website with your photograph, record credits, availability, etc. Upload your demo, perhaps even parts of songs you have engineered. Check the legalities of this. Do not upload anything that has not yet been released.

As well, the computer is a great tool if you are learning to play a musical instrument. Use the computer to take on-line music lessons, to download song tablatures, and maybe even play along with other on-line learners.

• **Start a band.** The best way to get as musical as you can is to play music. No matter what level of player you are, there are others out there who are in the same boat. Even if you can only play three chords, start a band and learn some three chord songs. But don't expect to sell out the local arena, just play for the fun and experience. Playing in a band lets you concentrate on music, timing, tuning, and getting along with musicians.

Chapter Two

THE STUDIO SETUP

The best recordings come from properly designed spaces. Before any equipment is loaded into the room, before any equipment setup begins, you must determine the best ways to take advantage of the recording space. Some studios have dead areas and live areas, each with their individual functions. Some studios have adjustable wall panels that can liven or deaden areas of the room. You may decide to, for example, record horns in a live area of the room and an intimate acoustic guitar in a less live area.

Understanding the physical limitations and characteristics of the room helps you decide where each setup will sound best. But before placing any microphones or cables, a complete grasp of the 'decibel' is paramount.

The Decibel

The decibel (deci: 1/10 and bel: Alexander Graham Bell) keeps track of the wide changes in acoustic and electrical levels. Bell determined one decibel (dB) was the smallest change in volume a human could perceive, no matter what the level. The decibel is simply a comparison of two numbers: the signal level and a predetermined 0 reference. With human hearing, the 0 dB SPL reference is the threshold of hearing. Bell used the decibel to assess, among many other measurements, sound pressure level (dB SPL), voltages (dBu, dBv, and dBV), power levels (dBm) and changes in signal level (dB).

Figure 2.1 shows how every doubling of pressure level equals an increase of 6 dB, no matter what that pressure level is. For example, one guitar might

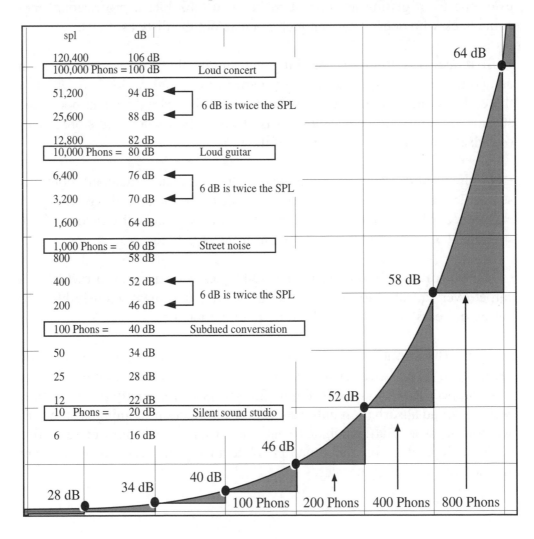

Figure 2.1 SPL vs. decibel

create a level of 80 dB. Bring in another one, and it doubles the SPL, or Phons, increasing the level by 6 dB resulting in 86 dB. If ten guitars created 100 dB, ten more would create 106 dB.

Room Preparation

• **Clean the place.** A clean studio, as with a clean control room, keeps everyone from getting anxious. It makes you look like a professional, not some hack who doesn't care enough to clean out the dirty ashtrays.

• **Use it or lose it.** Remove everything that isn't involved in your session, such as another client's equipment. Every item in the room rattles, so if it isn't there, that's one less rattle to worry about. As well, when the client books the studio, he wants the whole studio. It isn't fair to him when you store someone else's equipment in the corner of the room.

• **Go ahead and ask.** In an unfamiliar studio, ask a staff member about any standard instrument placements. Then use that information, along with your experience when placing each instrument. Maybe sketch the layout of the studio and use it to decide where each instrument will go.

• **Deal with noisy floors and chairs.** Oil squeaky chairs and throw down a carpet over squeaky floors. Everyday noises that are unconsciously blocked out by our ears are not blocked out by the microphones.

• **Diffuse the situation.** Use diffusion devices, such as sections of acoustic foam, stand-alone baffles, perhaps sections of plywood placed on the floor to fit your specific session needs. Figure 2.2 shows how properly placed baffles will block and absorb most initial reflections. Reflections coming off the wall are not an issue. Baffles might be used as absorption devices or reflective devices, depending on the situation. Note that some frequencies are not affected by absorption devices.

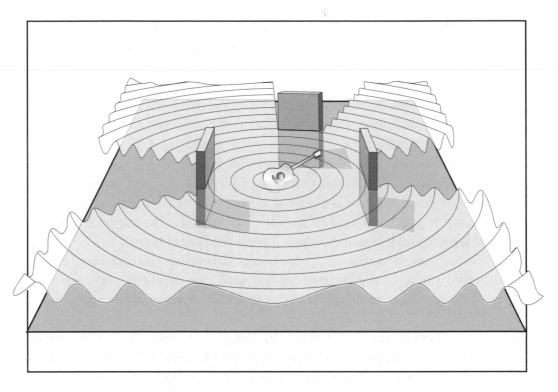

Figure 2.2 Baffles used for absorption

• **Any size recording studio can make hit records.** Good equipment means nothing unless the players are up to the task. Thinking you will get a great recording because you use the best gear is like thinking you can do a great painting because you have the best paintbrushes. Even a broken pencil can draw a masterpiece.

Placement

• **How many players are recording basic tracks?** Most recordings are broken down into three sections: basics, also called bed tracks, then overdubs and mixing. In today's studio, these overlap considerably. Determining how many players will be in the studio during the basics allows you to choose which and how many microphones to use on each setup. Different situations will require different approaches.

• **What is the style of music?** Determine player placement by using the style of music, how prominent each instrument will be, and how many players there are. Is it:

– A jingle? Jingles are advertisements heard on radio and television – often 30 or 60 seconds long. Recording will often be completed in a single day so there will not be a lot of time for experimentation. The most important factor here is usually the time restraints. Because the musicians that play on jingles are professionals, they will most likely play the part properly every time. The emphasis here is not on setup or performance. It is assumed that the setup is good and the performance will be fine. The important factors here are time and money. Set up and press the record button.

Proper documentation is key for a smooth running jingle session. Many different versions of a commercial, or 'spot' are recorded, and each spot needs proper documentation during recording, editing and mixing.

– A movie soundtrack? As movies get cheaper and more accessible, more people are getting into recording movie soundtracks. While most modern records are recorded to feature vocals or a certain instrument, movie soundtracks are less focused on the individual player and more on creating a feel that complements the movie. There is no strict standard, and you would record what each scene requires, whether a large group of players or an individual musician. As well, you must deal with 'movie' people. To fit in with everybody, put your hair in a ponytail, wear shoes with no socks, and call everyone 'babe.'

– A power rock band? They might sound better in a larger live space. Start the session by concentrating on bass and drums, but also record reference vocals and guitars. All aspects of the recording, production and playing must be of the highest quality. The world (hopefully) is going to hear the results, so take the time to set up properly and record everything correctly.

– A jazz session? You may prefer a mellower, dryer feel from a less reverberant space. With jazz, often players will want to watch each other without headphones. Everything is recorded live with no overdubs. Have enough tape on hand because left to their own devices, jazz players will never stop playing. Sleep? Food? Time itself? All secondary to the groove.

— A demo? Recording demos is not the time to spend four hours getting that perfect guitar sound. Take less time on setup, and more time recording.

— A television show? Many televisions have smaller speakers that cannot recreate a wide range of frequencies. Therefore, a piano, for example, recorded for a television soundtrack might not have the same equalizer setting as a piano recording for a rock record because on a television sound-track, boosting low, low frequencies may not have much of an effect.

— An independent punk band? They might sound best in a smaller, tighter 'garage' style room. Not a lot of setup time, and often raw sounds with live vocals.

— A solo acoustic player/singer? Setup may sound best in a smaller more intimate space, and the voice and the musical instrument might have separate microphones. Maybe the microphones will be placed in stereo in front of her.

• **Places please.** Most musicians, if they have been playing together for a long time, get used to certain placement among the rest of the players. Maybe the rhythm guitar player prefers to stand on the left side, while the bassist may like the right. Ask the players if they are used to playing in specific locations in reference to each other.

• **Don't isolate the players.** Just because an amplifier is isolated doesn't mean that the player must be isolated as well. Commonly, players being recorded together should be able to see each other, and at least one should be able to see the engineer. The person who can see the engineer normally gets a talkback microphone.

Microphone Stands

• **Don't use a large boomstand.** Sometimes a smaller boomstand will be less obtrusive. Microphone stands can range from a small stand to hold a tiny microphone, all the way up to a large boom with a big microphone hanging off the end of it. Match the stand size with the situation.

   Microphone stands touch nothing but the floor. A microphone stand won't tip over unless it was teetering in the first place.

• **Use a large boomstand.** Larger boomstands tend to be more stable, with less resonances and rattles, and they help eliminate rumble from the floor.

• **Duct tape rules.** Leave a couple of rolls of duct tape around, but only use it to stop resonance and rattles, not to hold a microphone stand. Duct tape won't hold a microphone stand in place for long, and it looks unprofessional. But sometimes there is no choice. If you must use duct tape to hold anything, don't wrap a whole roll around the microphone stand. A few times around will do.

• **Stand up.** Place all stands and run all cables before attaching the microphone to the stand. Wheeling large boomstands around a cluttered room is just asking for damage. Set microphones up last, and break them down first.

• **Secure the boomstand.** If you must use a smaller tripod boomstand, place the stand so one leg is directly under the boom, then sandbag the other two legs to keep them secure. Place a sandbag on the base of the stand to ensure stability. If someone bumps into it and it topples over, it's your fault.

• **Pair of matching shocks.** When setting up to record something in stereo, use matching stands and microphones. This not only looks professional, but makes both microphones sound alike. If one microphone has a shockmount, and the other doesn't, a slight rumble may creep into the one without the shockmount. Then you might need to add a low end roll-off, thus changing the intended matching sound.

• **Weight for the boom.** Set the counterweight of larger boomstands high enough so no one hits their head. If someone gets smacked in the nose, it gives the studio a black eye.

Microphones

• **What is a microphone?** A microphone is a transducer that changes acoustic energy to electrical energy. Sound causes the diaphragm within the microphone to vibrate, creating a small voltage. Louder signals cause the microphone's diaphragm to vibrate more, creating more voltage.

The diaphragm is the internal membrane (or the ribbon – depending on the microphone) within the capsule that vibrates. Different microphones will display different frequency response charts.

A small diaphragm microphone will commonly, but not always, accentuate mid to high frequencies. Due to the smaller mass of the smaller diaphragm, the low frequencies may not be as accurate as in a large diaphragm microphone.

Large diaphragm microphones might offer the widest range of frequency response to capture warmth, as in low strings or vocals. These microphones tend to naturally recreate better low frequencies, simply due to the mass of the diaphragm. The larger the mass, the less accurate the transient response.

• **Do a listening test.** When you understand the characteristics of each microphone, you are better equipped to choose the best microphone for each situation. Listening tests such as this will help train your ears to determine which microphone does what.

Set up the best few microphones available, perhaps four or five at a time, with the individual capsules adjacent to one another. Use no pads or roll-offs, and match the polar patterns. Set the faders so all the microphones have the same perceived volume. Have someone stand at an equal distance from all microphones and speak or sing. Turn one microphone on at a time and check which ones:

– Have nice crisp high end.

– Have a warmer midrange.

– Retain warmth and smoothness no matter what frequencies are altered. Lower quality microphones may start to sound brittle as you boost the higher frequencies.

– Have a thick bottom.

– Have hums or buzzes. Microphones with inherent lower levels may have more buzz simply because the level must be raised to match the rest of the microphones being tested. Older tube microphones can have this problem but their great sound makes up for it.

– Change level uniformly as the person moves about the microphone. Some cardioid microphones are more omni-directional than others, and some may not be omni-directional at all frequencies.

• **What is the proximity effect?** The proximity effect is the increase in a microphone's low frequency response when it's placed very close to a sound source. Microphones with omni-directional patterns are not affected by the proximity effect.

• **What is phantom power?** Condenser microphone capsules are very high impedance, so they need an impedance-converter circuit that requires power to operate. The power is called phantom power, and it comes from each mixer microphone input. Dynamic microphones have no active electronics, so phantom power is not needed.

• **What is a pre-amplifier?** Whether inside or outside the console, the microphone pre-amp raises the microphone signal level to a usable 'line level.'

• **What does frequency response mean?** A device's frequency response is its output level versus frequency. Different microphones will display different frequency response charts. For example, some microphones will have a smooth high end, while others may have an upper midrange bump.

• **What does transient response mean?** Transients are the initial sudden peaks of a sound, and are very short in duration. Transient response is the measure of how quickly a device responds to these transients. A percussion instrument contains very high transients, as does vocal sibilance.

• **Are you experienced?** With experience, you will learn the characteristics of each microphone, and choosing which ones are best for different situations will come naturally. Microphones are the tools of your trade, learn them.

Dynamic microphone

Dynamic moving coil microphones employ a coil of wires attached to a diaphragm, which is suspended within a magnetic field (Figure 2.3). Acoustical vibrations cause the diaphragm and the coil to vibrate within this magnetic field, creating current that electrically represents the audio signal.

Dynamic microphones tend to be robust, and are commonly used for the close miking of louder instrument setups including guitar amplifiers, drums and live vocals. Characteristics include a good transient response, a natural midrange peak (about 5 kHz), solid low-frequency response and tight polar pattern to keep leakage at bay. Because of their rugged build, dynamic microphones are often used in live situations.

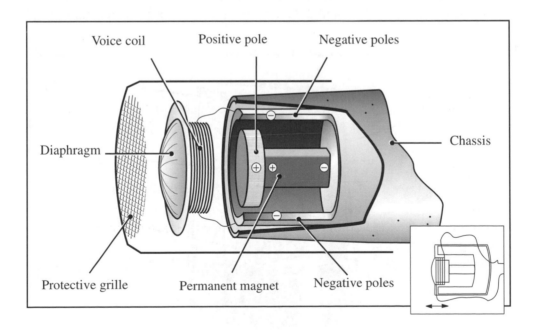

Figure 2.3 Dynamic microphone

Condenser microphone

Condenser microphones use two adjacent plates. One is stationary, while the other, a diaphragm, vibrates to incoming acoustic signal. These two plates are charged with a constant voltage – phantom power. As the distance between the diaphragm and the stationary plate varies with incoming vibrations, a varying electrical current is generated (Figure 2.4).

Condensers are less robust than dynamics, usually a bit warmer and smoother, and they capture a wide frequency range. They tend to be used on instrument setups such as vocals, acoustic instruments, room ambiance and less powerful electric instruments. Because all condenser microphones have a pre-amp, their output voltage is much higher than a dynamic microphone.

• **What are tube microphones?** Standard condenser microphones use internal transistors to pre-amplify the very small electrical current produced by the charged diaphragm. Some condensers use tubes. These tube microphones have their own power supply that charges the tube and the diaphragm. Tube microphones are expensive and delicate, but create a super warm round sound.

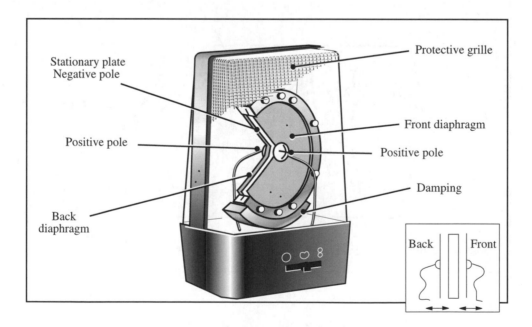

Figure 2.4 Condenser microphone

Ribbon microphone

Ribbon microphones use the same principle as dynamic microphones, only rather than a coil of wire, they use a ribbon of metal foil (Figure 2.5). This foil ribbon vibrates within a strong magnetic field creating a small electric current. This field is strong. Not strong enough to pull your belt off, but maybe strong enough to erase a tape, if left too close. Keep your tapes away from ribbon microphones.

Ribbon microphones traditionally record wonderfully lush low end. Great for male vocals or bass. Be warned – these microphones are quite sensitive to air movement and are not as robust as dynamic moving coil microphones; too much signal can stretch the ribbon. As well, ribbon microphones have a degree of natural internal compression.

Due to the sensitive nature of the ribbon, make sure the phantom power is switched off when using a ribbon microphone.

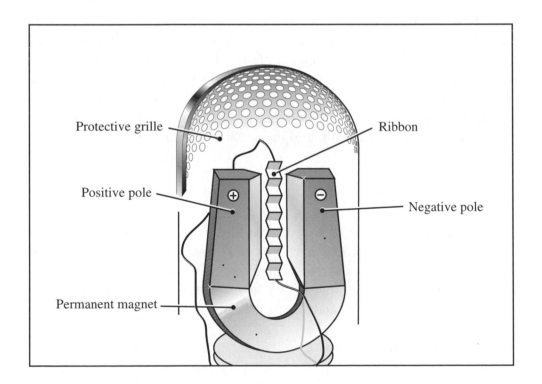

Figure 2.5 Ribbon microphone

Additional microphones

Additional microphone styles include:

Boundary microphone. A boundary microphone uses a flat plate and adjacent microphone capsule placed, commonly duct-taped, on a wall or floor to pick up direct and reflected sounds in phase.

Lavalier microphone. Contact or lapel microphones are small and commonly used for unobtrusive recording, such as when clipped to a lapel in television interviews.

Shotgun microphone. This ultra-directional microphone receives all the incoming sounds, but cancels out everything except for what is directly in front of the microphone.

• **What are pads/roll-offs?** A microphone pad is a circuit inserted at the microphone to lower the level reaching the console, such as -10 dB, -20 dB, even -30 dB. A roll-off is a low-end or high-end filter on the microphone used to eliminate the high or low frequencies that are not needed as an essential part of a sound. Use a low-end roll-off (HPF or high-pass filter) to minimize rumble on a vocal track, or a high-end roll-off (LPF or low-pass filter) to reduce sizzle from a high hat microphone without adding equalization.

Polar Patterns

A polar pattern is a graph of microphone sensitivity versus the angle of the incoming sound. Signals picked up in front of the microphone are called on-axis. Signals arriving away from the front are called off-axis. Some microphones have only one pattern, and others have switchable patterns, and even others have changeable capsules, each containing a different pad or polar pattern.

• **What is a cardioid pattern?** Cardioid patterns are so called because their pickup pattern is shaped not unlike a heart (Figure 2.6). Note that pickup on the front is best, and level decreases as sound arrives from the sides and back. As frequencies lower, cardioid patterns become more omni-directional.

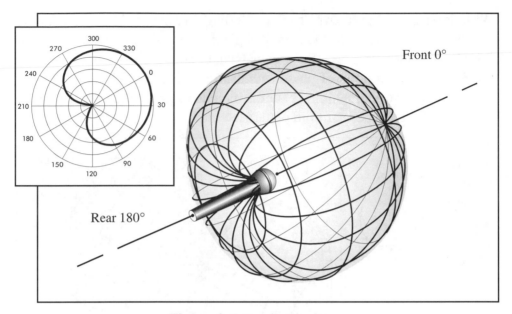

Figure 2.6 *Cardioid pattern*

• **What is an omni-directional pattern?** Omni-directional patterns are so called because they pick up signal from all sides equally (Figure 2.7). As the frequencies rise, the omni microphone becomes more directional.

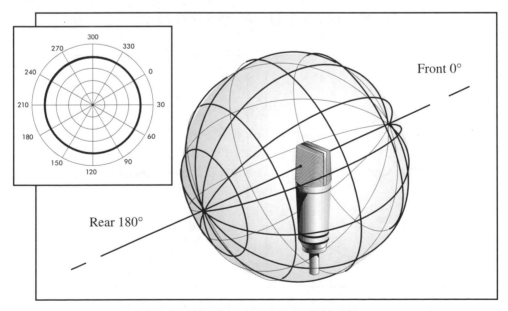

Figure 2.7 *Omni-directional pattern*

• **What is a figure-8 pattern?** The figure-8 (bi-directional or bi-polar) pattern is so called because it is shaped like an 8, picking up signal from both sides of the microphone while rejecting off-axis signals. In bi-polar patterns, one side of the pattern is internally set out-of-phase with the other (Figure 2.8).

Additional, more directional patterns include Figure 2.9(a) super-cardioid and Figure 2.9(b) hyper-cardioid patterns.

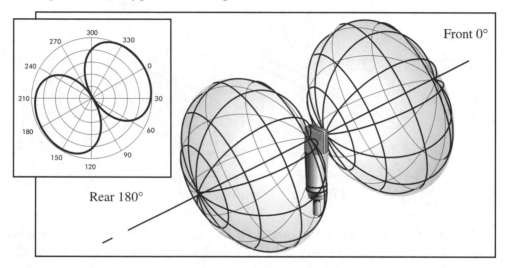

Figure 2.8 figure-8/bi-polar pattern

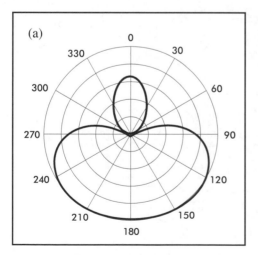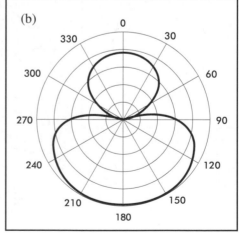

Figure 2.9 Super-cardioid (a) and hyper-cardioid (b) patterns

Microphone Choice

• **Match the microphone to the situation.** Match the characteristics of the microphone with the characteristics of the instrument to complement its sound. If, for example, a microphone has a natural low-end boost, maybe use it on a small acoustic guitar that needs a low-end boost, rather than on a bass amplifier, which may have enough low frequencies. Questions to ask before choosing the microphones might include:

– Is a microphone even necessary? Some instruments have a direct output that can be used instead of a microphone. This means no leakage from other instruments. Sometimes, such as in a live situation, the best microphone choice is no microphone at all.

– How much air is being pushed? Does the situation call for a dynamic, a condenser, a ribbon or another kind of microphone?

– What is the dynamic range of the musical instrument? A piano has a wide frequency range, a triangle doesn't. Instruments with narrow frequency ranges may not need the best microphone in the house.

– Will an older tube microphone work better than a new one?

– Should a small or large diaphragm microphone be used? Different microphones naturally produce different responses – the ears are the final judge. Commonly, lower-frequency instruments benefit from large diaphragm microphones due to the larger diaphragm mass which improves low-frequency response.

– How loud is the microphone? A microphone with a high output might not be the best choice for a really loud instrument just like a low-sensitivity microphone might not be the best choice for a low-output instrument.

– Where is the best placement for optimum response? Does the situation call for direct signal, close miking or not so close miking?

– What microphone polar patterns are available, and which is best for this situation?

– How many microphones are needed? Will one be enough, or are two (or more) needed? Should stereo miking be used?

– Is a pad or roll-off necessary? After the levels have been lowered, is the signal still too loud?

– Should I have the chicken or the fish for lunch?

Microphone Placement

• **Treat microphones with the utmost care.** Carry them one at a time, and don't place them on the floor. Bring them in and set them up. If you must put them down, use a blanket or towel. Older tube microphones are expensive to repair and repair can take months.

• **Try all the microphones available.** Don't forget about that one humble microphone that sits in the corner of your arsenal for years and never gets used. I wonder how a shotgun microphone aimed at a guitar amplifier would sound, or a boundary microphone taped to the floor under a guitar amplifier, or a lavalier clipped to the hole of an acoustic guitar?

• **Check please.** Double check, then triple check that the microphone is attached to the stand correctly, not touching any other stands or instruments, and is solidly in place.

• **Mute the microphone before plugging in the cable.** At this point of your setup, the proper cable should lie at the base of the microphone stand. Check with the control room that the channel is turned off. This avoids a great loud pop in the speakers and, even worse, through all the headphones.

• **Turn off the phantom power.** If the microphone has its own power supply, as many tube microphones do, turn off the phantom power at the console. As well, turn it off for dynamic and unbalanced microphones.

• **Three-to-one rule.** There is an old rule when placing two or more microphones on a single source. Place the second microphone at least three times the distance between the first microphone and the sound source. At this distance, the second microphone picks up enough of the room ambiance to minimize any phase interference between the two microphones. Figure 2.10 shows that microphone (a) is one foot from the acoustic guitar source, and that microphone (b) is beyond the three-foot perimeter.

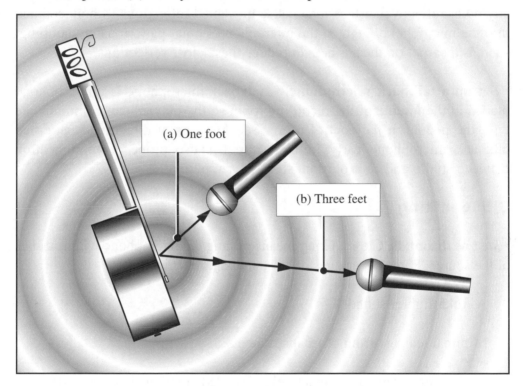

Figure 2.10 Three-to-one rule

This rule can be broken when two microphones are placed so close together that they are in phase, such as two microphones taped together and aimed at a snare drum.

• **Four-to-one rule.** The distance between the close and distant microphones is far more critical if the close microphone is very near the sound source. When miking an acoustic guitar with the close microphone at about three inches from the fret board, the distant microphone should be at least four times that distance. When close miking in a somewhat dead acoustic space, the four-to-one rule should be applied. In a more traditional open miking situation, the three-to-one rule can be applied.

• **Port noise complaint.** Blocking the ports of a microphone will change the polar pattern. Placing a microphone too close to the floor, wall, window or any other reflective surface may cause unwanted reflections to bounce back into the microphone. Tighten the microphone's polar pattern using damping, such as a carpet on the surface, to reduce these close reflections.

• **Set up a talkback microphone.** A talkback microphone is a microphone placed in the middle of the recording room so the players can communicate with the people in the control room. Set up an omni-directional patterned talkback microphone early in the setup because proper communication between studio and control room shortens setup time.

Close miking

• **Just as it sounds.** Close miking means placing the microphone within inches of the source. Some advantages of close miking might include:

– A fuller tighter sound.

– Minimal leakage from other instruments.

– No unwanted ambiance on a track. If needed, ambiance can be added to a dry track, but once recorded, ambiance cannot be removed.

– Easier to return to the same sound if ambiance is different, such as when using a different studio.

– More separation when recording in stereo.

• **Fine placement.** You can really hear differences in the sound as you move the close microphone around in front of the source. As the microphone moves away from the source, slight changes in microphone placement aren't as apparent.

A microphone placed too close to a source can compromise warmth. The hottest part of the match is not right at the head, it is farther away in the flame. The strongest part of a waterfall is not at the crest, but farther down, where it has had time to build power.

Distant miking

• **Go the distance.** Distant miking means the microphone is placed a few feet or more from the source. Advantages of distant miking include:

– A more ambient sound with greater influence from the surrounding environment. Ambiance can help establish a sense of depth. Take note: ambiance microphones, also called room microphones, can reveal flaws in the room design.

– Not in the way of players, as they need to be totally comfortable.

– A second distant microphone recorded on a separate track can add placement flexibility when mixing.

– Able to pick up a large group of musicians. (Have you ever tried to pick up a large group of musicians?) When distant miking a group of players, usually for classical music, the tonal balance captured can be better than recording with separate close microphones. The players can mix themselves better than the engineer.

– Less dynamics, often associated with, for example, vocals. With close miking, a compressor is almost a necessity, unless the singer is really good and knows how to control the dynamics of his voice. A foot or more distance between singer and microphone causes less change in level. The farther away the singer is from the microphone the more the surrounding environment gets recorded with the vocal. This is when studio design comes into factor.

X/Y stereo miking

To produce a full stereo sound, place matching microphones (often microphones that have concurrent serial numbers are very close in sound) in an 'X' position with their capsules close together (Figure 2.11). This setup essentially eliminates any phase problems between the two microphones. Use X/Y miking in restricted spaces where microphones need to be somewhat close together yet produce a good stereo image such as over a piano, or above a violin section.

The angle of the microphones will determine the stereo spread. The wider the angle, the more stereo spread, until the angle gets too wide, creating a hole in the middle. The narrower the angle, the less of a hole, creating more of a mono signal.

Set up the two microphones so both are at the same height and have the same directional polar pattern with matching pads and roll-offs.

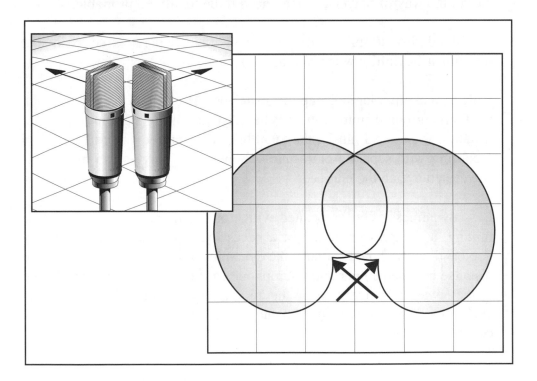

Figure 2.11 X/Y microphone setup

'Spaced pair' stereo miking

A standard spaced pair technique places the microphones far apart within the recording room. The darker areas on Figure 2.12 show the common pickup areas. Setups might include:

(a) Quite a few yards apart. These room microphones pick up an ambient stereo room sound.

(b) Two feet to three feet apart. The wider the spacing, the wider the stereo spread. This setup gives a good stereo image without too much ambiance.

(c) Parallel, in front of an instrument. Both microphones parallel, and aimed toward the source gives a wide stereo image. For an even wider image, place a baffle between the microphones.

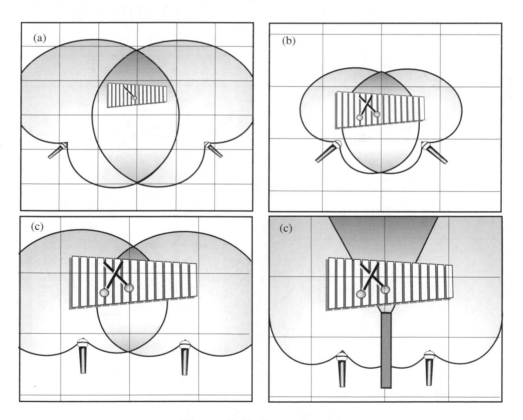

Figure 2.12 Spaced pairs

M/S stereo miking

M/S stands for mid-side. This method sums and differences the polar patterns of two microphones. Use M/S when you need to decide later how wide the stereo image should be (Figure 2.13). To monitor and control the depth of the stereo image, three channels are needed on the console.

(1) Set up two microphones, one with a cardioid pattern 'M,' the other with a figure-8 pattern 'S.'

(2) Aim the M microphone toward the source. Position the S microphone on the same plane, but at a 90 degree angle.

(3) Route the M microphone to record on track one, and route the S microphone to record on track two. The M microphone returns from the multitrack recorder into channel one, and the S microphone returns into channel two.

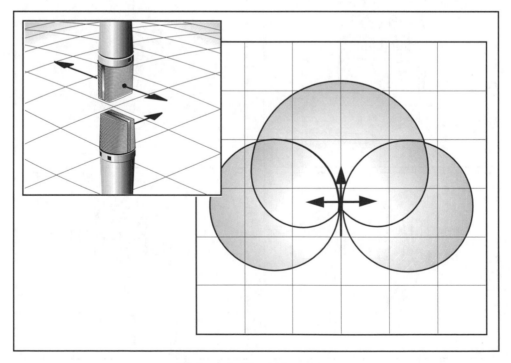

Figure 2.13 *M/S pattern*

(4) Patch the S signal out of channel two into channel three so it returns on both two and three. Switch channel three out of phase.

(5) Pan channel one to the center, then pan two and three hard L/R.

(6) Raise channel one to set the appropriate mono level, then raise channels two and three to control the stereo width.

Decca tree stereo miking

The decca tree setup, used for larger situations, uses a three microphone array (Figure 2.14). All three are set in an omni-directional pattern, then placed left, center, and right a few feet behind and about ten feet above the conductor. Decca tree recording is becoming more popular in L-C-R (left, center, right) stereo, and 5.1 surround DVD recordings.

Decca tree stereo miking records not only on the left/right location of instruments but the fore and background location, crucial for orchestral sessions where depth and dimension would be lost with close miking.

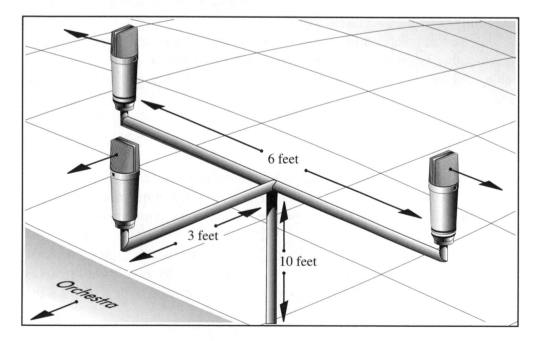

Figure 2.14 Decca tree

Cables

• **What are balanced cables?** A standard recording studio XLR balanced cable uses three wires – two to carry the signal and one for the shield. The three wires are termed hot, cold and common. Both XLR and 1/4" stereo cables are balanced. On a balanced 1/4" cable the three wires connected are tip, ring and sleeve (TRS) with the sleeve being the shield terminal.

Of the two ends that carry the signal, one has the polarity (phase) switched. Why? Any interference will affect both wires equally. At the balanced input connector, these equal and opposite interference signals will cancel out. This process is called 'balancing' or common mode rejection. This eliminates interference so you can use longer cables with minimal hum pickup.

• **What are unbalanced cables?** A standard guitar cord is an example of an unbalanced cable. It contains two wires – one wire is the hot, and the other is the shield. The shield is wrapped around the hot wire and used for the return signal. Unbalanced cables are normally used for getting mono signal from the instrument to the amplifier or direct box. Most outboard equipment uses 1/4" inputs, and all consoles have the option of balanced and unbalanced 1/4" (line level) inputs.

Another unbalanced line is a 1/4" non-shielded speaker cable. It is two heavy-gauge parallel wires. Don't use speaker cable in place of shielded cable. Cables from amplifier to speaker must be heavy gauge for as little power loss as possible.

Running cables

• **Use the proper cable and connectors.** Adapters will always degrade the signal so use them as a last resort. For the best results, use the proper cables so no adapters are necessary. One long cable is better than joining two short ones.

• **Respect the studio cables.** Wrap and unwrap studio cables correctly and don't throw them around. Avoid any sort of sticky tape to hold cables together. Use a specific plastic clip, even a piece of rope with a loop on the end, anything but tape.

• **Label the cable.** Label both ends of XLR cables for easier tracing. For example, if cable 28 is connected to a microphone that is not working, rather than having to trace it through the labyrinth of all the other cables, simply read the number on the cable, then check where the corresponding number is plugged in at the input panel.

• **Keep it short.** Use the shortest unbalanced high-impedance cables as possible. Longer cables can pick up hum and roll off high frequencies.

• **Keep it long.** Allow enough length so cables are never taut. Don't wrap them up in a coil, just leave the excess cable on the floor by the base of the microphone stand.

• **Leave a channel or two open when plugging in the drums.** You want all the drum microphones to enter the console in the same area. Later in the session, if you need to add another drum microphone for any reason, you won't need to return it somewhere else on the console.

• **Use the best cable.** Don't use a suspect cable on a microphone. If you're short on cables, use it on headphones. If it fails during the session, the recording will not be compromised. No one wants to find out half the piano sound is gone when it comes time to mix. Remove faulty cables from the room so no one mistakes them for usable cables.

• **For best results.** Use the shortest thickest highest quality cables on the vocal microphone.

• **Check your shorts.** Keep an ohm meter close by to determine where a problem lies. Is there a short in the cable? Is the microphone not working? Maybe it can be as simple as a wrong button pressed at the console. A meter tells you instantly if signal is correctly flowing through a piece of gear.

• **Three cord rock and roll.** Keep extra everything close by. The session should never stop due to a lack of workable cables.

• **Power arrangers.** Don't run (high-level) AC power cables parallel with (low-level) audio signal cables. Cross power cables at right angles from signal cables to minimize AC buzz leaking into the microphone cables.

• **The wall of sound.** To avoid interference, run cables a few inches away from the studio wall. Right behind this wall lies a plethora of power cables and other audio lines.

• **Don't plug everything into one AC outlet.** You will have a hot, smoking studio with scorching leads and blazing tracks that smolder! Get it?

• **Don't wrap the cable around the stand.** Unwrapping and removing a cable because the placement is wrong is a waste of valuable setup time. Sometimes, just leave it to hang.

• **Wrap the cable around the stand.** If you are sure the microphone stand is properly placed and the signal flow is correct, maybe wrap the cable around the stand to keep it out of the way.

Direct Box

A direct box, or 'D.I. box' changes an electric instrument's high-impedance unbalanced output signal to a low-impedance balanced signal that will properly interface with the microphone pre-amp. Some are designed with an input pad and a ground switch on them. Like microphones and amplifiers, they vary widely in sound. Direct boxes are available as passive (transformer only) or active (pre-amp and transformer) where amplification allows level to be raised higher to improve signal-to-noise ratio. Passive direct boxes might sound more natural than active ones due to lack of additional processing, but active boxes may have better frequency response.

Advantages of using a direct box include the absence of leakage, no amplifier distortion and a full clean sound. A good direct box records the full spectrum of the instrument, where a microphone in front an amplifier may not. Many engineers agree that a direct box alone just can't recreate that chest-cavity vibrating thump that an amplifier and microphone can deliver.

Ground

Ground loop and the resulting hum occur when electrical equipment is grounded to more than one location, creating multiple paths. As well, hum can originate from fluorescent lights, video monitors, dimmer switches, refrigerators and more.

• **Eliminate the hum.** If your audio equipment setup is humming:

– Use a dedicated AC power circuit for the studio. Some studios have outlets strictly for audio equipment, often indicated by orange wall outlets. Sometimes all the audio equipment is plugged into a large 12-outlet box, with everything on the same circuit. The outlets in the rest of the studio would be on a separate circuit, and used for the coffee machine, the video games and the vibrating bed.

– Don't plug a power strip into another power strip.

– Eliminate all fluorescent lights in the vicinity.

– Don't use a ground lifter – a converter that changes a three-prong plug to a two-prong plug. This can be dangerous. The third pin is there to keep you safe. Most equipment today is fitted with three-prong outlets, the third being a chassis ground just to avoid any bothersome recording engineer electrocution issues.

– Use a transformer, or hum remover, made specifically to eliminate hum by removing any common ground between devices.

– Use the highest quality, shortest cables.

– Rearrange the way the outboard equipment is stacked. A transformer in one unit may cause hum in an adjacent unit.

– Use a filter. Although not recommended, a 60 Hz hum can be removed using a filter set at 60 Hz and 120 Hz. But this pulls that frequency on any musical content and can affect harmonics.

– Switch the direct box ground. Some direct boxes have a ground switch which disconnects the input cable shield from the output cable shield, eliminating ground loops and the resulting hum.

– When the hum is from a guitar amplifier, plug the amplifier into a different wall outlet on a different circuit.

– Have the player rotate. Single-coil pickups in guitars are notorious for humming when the player is aimed in a certain direction. The hum disappears when he points in another direction. When you find the quietest angle for the player to stand, run a section of sticky tape across the floor so the player knows just what the angle should be.

– If you still can't find the hum, use the process of elimination: shut everything off, and start at the beginning turning everything on again. When the hum starts, that's the culprit.

Phase

When identical signals combine, such as when two microphones are bussed to one track, phase shift can be introduced when one of the signals is delayed. This shift or delay is measured in degrees. Figure 2.15(a) shows that when similar sine waves combine with no delays, the signal doubles. Figure 2.15(b) shows that when one signal is delayed exactly half the length of the sound wave, its polarity is 180° out-of-phase with the other signal. When combined, these signals will cancel each other out. Note that phase or phase shift involves delay. Polarity does not. Polarity refers to the + or - direction of an electrical signal.

• **Check for polarity.** Somewhere along the line between microphone and monitor, the polarity may be reversed. Perhaps it's a miswired connector or input jack, maybe even a wrong button pressed. To check for polarity:

(1) Place matching microphones right next to each other, with the capsules as close as possible, both aimed toward a source.

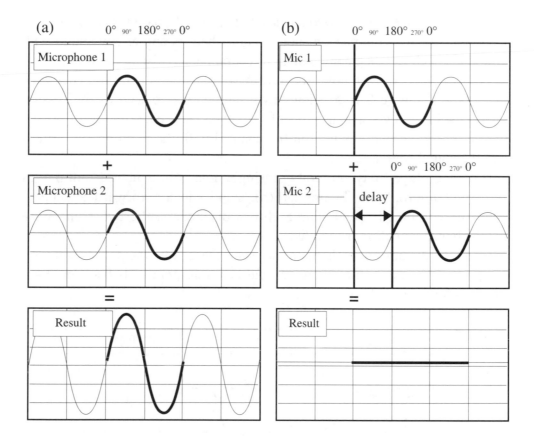

Figure 2.15 *Phase cancellation*

(2) Have someone stand a few feet away and speak.

(3) Solo the two channels, and listen in mono.

(4) Press the phase button (sometimes mislabeled as polarity) on one of the microphone channels. If the signal cancels, the microphones have the same polarity. If the signal cancels when no buttons are pressed, the two microphones are wired in opposite polarity (the wiring to pins 2 and 3 are reversed in one microphone's XLR connector).

• **Phase – the facts.** Every foot that you move a microphone away from the source results in about a 1 millisecond delay. If a second microphone is placed incorrectly, the signal reaches it at a different time in the cycle.

At specific distances, certain frequencies will be out-of-phase resulting in a thin incomplete sound. This phenomenon occurs when recording any sound with two inputs, such as when combining a direct box and a distant microphone.

Figure 2.16 shows how to minimize phase cancellation between microphones. Figure 2.16(a) shows two microphones placed as the session dictates. Solo the two channels, match the levels, and listen in mono. At the console, press the phase button on the channel with the microphone farthest from the sound source.

Have the player play, preferably a single note, while an assistant moves the distant microphone, Figure 2.16(b), stand and all, forward and back. When the combined sound source almost disappears, the two signals are out-of-phase, canceling each other.

Place the distant microphone in that spot. Deactivate the phase button on the console. Figure 2.16(c) shows the farthest microphone is back in phase with the close microphone.

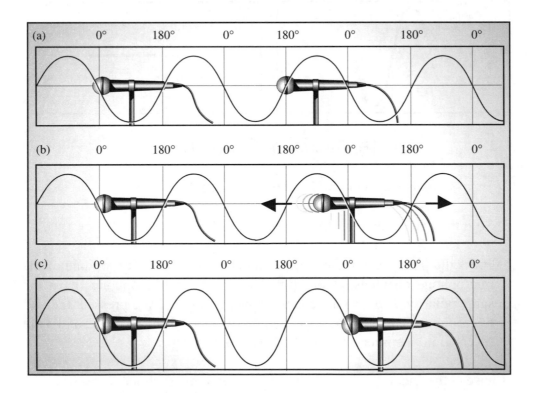

Figure 2.16 Minimizing phase cancellation

Chapter Three

THE DRUMS SETUP

There is no great trick to making a set of drums sound like a well tuned, punchy, solid drum set. First, start with a well tuned, punchy, solid drum set. Properly tuned, well played drums coupled with good songwriting and solid engineering will always record the best.

But let's get real. Some drummers aren't interested in changing heads or tuning. They just want to bash away. If this works for the music and the artist, then bask in the bash.

Placing the Drums

• **Arrive early.** If you are new to the room, or new to the drummer, set up a bit earlier than normal, perhaps even the night before the session. This should give you and the drummer extra time to work out the absolute best sounds. But don't tire him out before the session starts.

• **Determine the best placement for the drums.** For a rock record you might place the drums in the middle of the room to minimize coloration from close reflections. Maybe place the drums next to a door that opens up into a larger area such as a loading bay, for ambiance microphones. This isn't necessarily for a huge ambient rock drum sound, but for using a bit of the room sound for placement and focus of the drums.

For a blues session, you might not place them right in the middle, but more to one side to leave space for other players. For a country session, maybe place the drums in an isolation booth. Ask the staff where other engineers normally place the drums.

Work with the drummer to find the best location. You don't want to place the drums in the isolation booth, then find out he wants to be on the floor with the rest of the band. Change your setup for him, don't expect him to change his setup for you.

• **Get felt in the corner.** Take advantage of the room's acoustics. Some studios have reversible panels on the walls where one side is reflective wood and the other side is absorbent padding. Use these acoustic panels to change the characteristics of different parts of the room. If these measures are not enough to eliminate errant frequencies or reflections, hang blankets and place baffles. Wait until all the microphones are set up before placing the baffles.

• **Rise and shine.** If a whole band is playing together, maybe place the drum kit on a drum riser to eliminate low-end rumble from neighboring amplifiers, and to keep rumble from the drums out of the amplifier microphones. The drummer might play a little better when he is at eye level with the rest of the players and the engineer. It makes him feel as if he's on stage.

• **Use a carpet under the drums.** Lay out a carpet before setting up the drum kit to keep reflections from bouncing off the floor back into the microphones. A carpet helps eliminate some of the squeaks and rattles that may occur when the kit sits on the bare floor. A carpet also helps keep the kit stationary.

But how does it sound without the carpet underneath the kit? Maybe a drum sound with reflections from the floor will work for your situation. Just don't scratch the floor.

• **His rug is taped on.** If you know the drummer is in again next week, tape the markings on the carpet under the kit. Use duct tape to mark exactly where each drum and stand goes. Different drummers using the same carpet may need different colors to distinguish whose kit goes where.

• **Most drummers will set up the kit themselves.** Once the drums are placed properly according the drummer, set all microphone stands squarely on the floor at the correct spot. Use the strongest stands available. The placement of the drums takes precedent over the placement of the microphone stands. Work the setup around the drums.

• **Use a sand bag to hold a stand in place.** Set all stands stable enough for someone to bump into it and have the placement remain unchanged.

• **Stop the kick from moving forward.** Some drummers really hit the drums, causing the kick to edge forward, shifting the whole kit. Secure the kick in place using a cinder block, sandbag or equivalent. Some situations call for nailing a short plank in front of the kick drum to keep it in place, but this is an extreme measure.

• **Fix faulty felts.** Replace worn felt pads and plastic sleeves that hold the cymbal in place. A worn cymbal sleeve may rattle against its metal stand.

• **Bring it from home to me.** Encourage the drummer, as with the rest of the musicians, to bring all of their percussion instruments to the sessions. Different songs may require different sounds. No one wants to hear, ' I have a tambourine, but I left it at home' or worse 'Oops, I just broke my last drumstick.'

• **Don't compromise the drum sound.** Take the time to fix problems before they reach the console.

Changing Drum Heads

• **More than one way to skin a kit.** As with guitar strings, drum heads (also called skins) wear out with use. New drum heads record much better than dead drum heads.

• **Do not change the drum heads without approval.** It's not your sound, it's the drummers. Some drummers prefer the sound of a dead snare, or certain tom-tom tunings. Don't monkey with any musical gear unless you own it, or you have approval. Musicians tend to get choked when they walk into the room and find their instrument disassembled across the floor, with you trying to somehow 'improve' the sound.

If you do change the heads on the drums, either replace them with similar heads, or find out what the drummer wants.

• **Go stand on your head.** If you must change a drum head:

(1) Use a proper drum key to remove the lugnuts.

(2) Remove the old head from the snare.

(3) Clean out any waste, and wipe the rim clean with a rag.

(4) Check for warpage on the rim by laying the drum upside down on a flat surface, turning out the lights and shining a flashlight inside the drum to check for any gaps. If gaps exist, the drum can't be properly tuned. It needs to be looked at by a competent professional drum dude.

(5) Seat the new head and connect each lugnut by hand.

(6) Tighten the lugnuts like a you would when changing a tire – not one after the other, but one then it's opposite across the drum – for example, one o'clock, then seven o'clock and on. Twist each lugnut once, then twist the next one once with single twist. Turn each lugnut gradually so the head tightens to the shell evenly. As the head tightens, lessen the amount of each turn. Expect cracks and crackle sounds. That is the head stretching. Press down on the drum as you tighten the lugnuts to really stretch the head across the rim. Sometimes drummers kneel, or even stand on the snare drum head to stretch it.

• **Front line assembly.** Removing the bottom head of the tom-toms or the front head of a kick drum results in uneven resonance in the drum shell.

Better to install an old head, and trim most of it out, leaving an inch or two around the perimeter. This leaves the front hardware intact, yet allows full access to inside the drum for a microphone, if that's your setup.

Tuning the Drums

• **You can't beat our drums.** If you took two wooden sticks and pounded on a guitar for hours on end, do you think it may need to be tuned now and again? It's understandable that pounding on a set of drums over and over, day after day tends to render the ever-loving life out of them. Keeping drums in tune with new heads keeps them musical, therefore easier to record.

Tuning works well when the drum heads are left to stretch for at least a day. Of course, this might not be an option when you are scheduled to record the drums right now.

• **Tune that thing up.** Proper tuning enhances the drums' overtones, and brings out fundamental resonance. Tuning the drums slightly lower produces a larger sound because more air is moved. Set too low, and the tone suffers. Tuning a drum higher raises the drums' natural lower frequencies. Tom-toms tuned too high may create additional rattle on the snares under the snare drum.

• **Turn the snares off before tuning the snare drum.** Tap around the outer edge of the drum head and listen for differences in pitch. Tighten or loosen as necessary. This takes a trained ear.

A tighter head gives more attack and crack, and a looser head gives a deeper, flappier sound with less crack. Some drummers leave the snares loose because they like the loose, rattle-like sound.

• **Tune that thing down.** Down tune a small snare drum for a massive snare sound. Some of those great big drum sounds you heard in music as a kid were not made by the huge kits you may have imagined.

• **Tuning two headed tom-toms.** With two-headed tom-toms, sometimes the drummer prefers the top skin tuned higher than the bottom. He'll tune the top skin slightly high, and the bottom skin slightly low to give the drums a more sustained and pleasing sound. Sometimes for a dryer sound he'll prefer the bottom skin tuned higher than the top. For a nice, full sound he likes the top head tuned the same as the bottom.

Bottom heads help control the tuning and decay of the drum, and give more body to the tom-tom sound. Two headed tom-toms can be harder to record than single headed toms, as the tuning can take longer.

Tune each drum, starting with the top head, then adjusting the bottom head. Keep going down until all tom-toms are tuned. Some drummers tune their drums by knocking on the shell or tapping the head of the drum, then listening to the resonant frequency and tuning the heads to that.

• **Intervals.** A common interval between tom-toms is fourths and, as some drummers say, a fifth between the lowest tom and the kick. Determining the interval tends to be the drummer's domain, not the engineer's. If he wants thirds, thirds it is. Ideally, each drum would be tuned to help underscore the musical chords of the song.

Smaller drums, cymbals, and percussion instruments such as tambourines and castanets are classed as indefinite pitch, or not of a fixed pitch so they aren't normally tuned.

• **Alter the pitch a bit.** Keep an even tension on all lugnuts so unwanted overtones are minimized. Maybe create a more complex tom-tom sound by loosening one lugnut to slightly lower its pitch.

• **Carr talk.** Well-tuned drums sound great, so take the time needed. When I worked with KISS, the late, great Eric Carr had a drum tuner in the control room. Eric would pound the drums for a couple of takes, then the drum tuner would tweak the tuning before the next take. The drums were always right.

• **Drummers always know best.** There is no right or wrong way to set up an instrument. Different people prefer different setups and the drummer knows best about his own drums. A good drummer will be able to hear the proper pitch, and tune the drums to minimize sympathetic ringing.

Preparation

• **Dampin' donuts.** To dampen a drum, place a donut (an old drum head with the middle cut out) on top of the drum head. The drum will have less sustain than a drum with no damping. For real damping, roll up a wad or two of duct tape, then duct tape these wads to the drum head. This renders it dead. Don't put anything over the area where the drumstick hits the head.

Commercial sticky gum is now available that tears off. Tear off a blob and smear it across the drum for damping. Some drum heads are manufactured with a damping system installed, such as a thicker outer rim.

Before damping, consider the song and the rest of the instruments. A full, fast song may not need drum damping. A drum sound with all the overtones might not sound as big and ringy when the rest of the instruments are playing.

• **King of the ring.** Work with the player to eliminate all the drum kit's squeaks and rattles – not to be confused with the natural ring of the drums. Harmonics are part of the sound of the drum kit. Many drummers like this ringing overtone, which is more a function of tuning. Some engineers, not understanding that this can be a desirable trait, work on eliminating it.

• **Listen, shove it.** To deaden the kick drum sound, place a folded packing blanket or a sand bag inside the kick. Maybe fill the kick drum with torn up newspaper, or light blankets and towels. Whatever works for the situation. But before shoving in a blanket, listen to kick drum. Maybe the sound you want is already there.

• **Will that be cash or credit card?** For a real click on the kick drum, glue a quarter right where the beater hits the head. If you are too cheap for that, tape a charge card (preferably the record producer's) to the kick drum head with duct tape. The beater creates a click when it hits the quarter, or the card, and not the drum head.

• **Wooden vs. felt.** A felt beater on the kick results in a softer, woofier sound. A wooden beater can give a clickier, defined kick sound. Many hard rock players use a wooden beater, while a lot of jazz players use felt beaters. These choices commonly fall into the drummer's jurisdiction.

• **We don't have much tape, so let's wrap it up.** Wrap masking tape or duct tape around the felt beater to give it more attack. Naturally, the tape wears out after a few takes.

• **Spring break.** If a new pedal spring is squeaking, oiling it may not be necessary, just try exercising it. Pull the spring farther ahead than it is supposed to go, but be careful not to break, or even damage the spring by overstretching it. There's nothing like a good stretch before getting started.

Miking the Drums

Miking the kick drum

• **Head space.** Different placements will result in vastly different kick drum sounds. The best way to find the perfect spot is to listen in the control room while your assistant moves the microphone around. As the drummer plays the kick drum, you listen for the best spot. When you hear it, tell the assistant to stop. That is the best starting point. Figure 3.1 shows a microphone:

(a) At close range. Aimed within a few inches from where the beater meets the skin, creating more click and attack than boom.

(b) Pulled back and off-axis. A microphone aimed towards where the shell meets the head results in a rounder sound, with definition, less attack and more boom. This is a common starting point.

(c) Distant. Pull the microphone to just outside the open head for a full deep drum sound, but watch for increased leakage and decreased clarity.

• **Dynamics.** Due to the high pressure of drums in modern music, dynamic microphones are commonly used on close-miking situations. Dynamics tend to be more robust and can handle the solid pressure levels. A large diaphragm dynamic microphone works well to accurately capture as much low end as possible. That kick can push a lot of power.

Note that some of today's condenser microphones are quite rugged, and certainly up to the task of recording the toughest of drums and drummers.

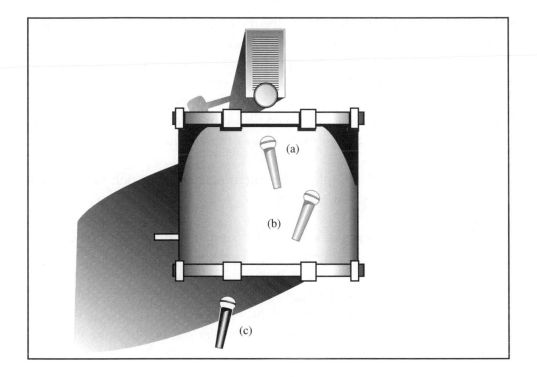

Figure 3.1 *Microphones inside the kick drum*

• **The style of music dictates placement.** For a heavier rock sound, you might put the microphone a few inches from the inner head, then baffle off the kick drum. For a more jazzy sound, you might leave the front bass drum head on, then place the microphone a short distance from the front head in a more open environment.

• **Embarrassing leakage.** Lessen leakage on the kick drum track by aiming the microphone inside the kick drum away from the low tom-tom.

• **Use a real bass drum.** If you aren't getting that round low kick drum sound that you desire, bring in a larger bass drum, even as large as a parade style bass drum, and set it close to the front of the kick drum. Set a microphone in front of the big drum. As the drummer plays, this low drum will ring sympathetically. Add this round warm low end into the sound as desired. Use this technique to give a smaller drum kit a more natural low end.

• **Build a kick drum tunnel.** Once the microphone setup is complete, build a tunnel around the kick drum. Figure 3.2 shows how this isolates the rest of the room from the kick drum microphone, and keeps the kick drum somewhat isolated from the rest of the room. Some engineers use a chair in front of the kick, with a blanket over it. Some use two microphone stands to hold the blanket. This setup works well with one close microphone, and a second one outside the shell aimed at the pedal. Follow this routine:

(1) Use a proper stand and place the microphone in the kick drum as you normally would to suit the music.

(2) Place two short microphone stands (or similar apparatus) in front of the kick drum, then lay a packing blanket over the stands, creating a tunnel.

(3) Duct tape the blanket to the hardware on the kick drum, not the shell. Ask the drummer before duct taping blankets to his beloved drums. For more isolation, cover the open end with another blanket, making sure you don't move the microphone.

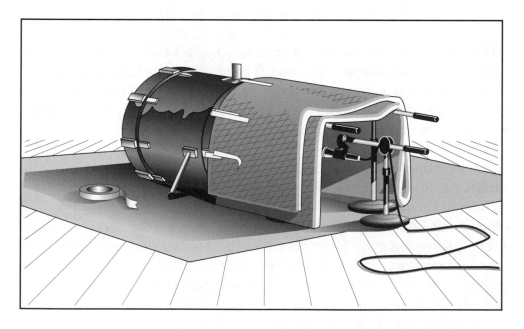

Figure 3.2 *Kick drum tunnel*

• **A last resort.** In a big session where microphone stands are at a premium, consider laying the microphone directly on the pillow inside the kick drum, aimed at the beater. Crude but effective.

• **A kick drum sound is only as good as the surrounding drum set.** Leave all the drums in the monitor mix when setting up the kick microphone. This lets you monitor frequency overlap between the tom-toms and the kick drum.

• **Let's dance on.** Don't underestimate the importance of a good kick drum sound. It carries the downbeat of the music. This is what people dance to.

Miking the snare drum

• **Start with a dynamic microphone.** A loud snare drum's high transients mean that a dynamic microphone may work best. Start by aiming the microphone across the drum head toward the center of the drum where the stick meets the head. Keep the microphone about an inch above the rim. Maybe aim the microphone off-center to eliminate some of the click and to coax more of the tonality from the drum. Listen and move the microphone to suit your needs.

   Some recording engineers aim the snare microphone off-axis to the high hat to minimize leakage.

• **Use more than one microphone.** Place a second microphone underneath the snare drum to capture the snares hitting the bottom skin. Since the bottom microphone 'sees' the heads move out when the top microphone 'sees' the heads move in, the microphone signals are in opposite polarity. Switch the polarity on the bottom microphone channel.

• **Head to head.** Tape a small, hefty condenser microphone along side of the dynamic microphone on the top of the snare drum. Position the capsules as close together as possible. This condenser microphone should accentuate the higher frequencies of the snare, while the dynamic microphone picks up the punch of the drum. Rather than equalizing the dynamic microphone to get more highs out of it, raise the level of the condenser microphone to add a nice crack to the snare sound.

If the situation calls for it, maybe use two different microphones on the top of the snare, plus one on the bottom. Of course this takes three channels on the console, more time, and may not even be necessary. Rarely will three microphones sound three times as good.

• **Eliminate room rattles.** Keep all unused snare drums covered, as the snares tend to rattle from the room sounds.

• **Wanted. Dead or alive.** When I worked on 'Bon Jovi – Slippery When Wet,' Tico Torres would remove pockmarks from his snare drum skin by slowly moving a lit portable lighter above them. The heat caused the pockmarks to recede, reviving an otherwise dead drum head for one more pass.

Miking the tom-toms

• **Dynamic microphones** work well on close miked tom-toms where the player hits hard. Condenser microphones sound good on less aggressive styles, as they capture the player's rich subtleties and dynamics. Some close-miked condensers may overload.

If possible, use the largest capsule microphones on the lowest tom-toms.

• **Single headed tom-tom.** If the tom-tom has a single head, place the microphone inside the drum. Aim it away from the cymbals to reduce leakage.

• **Bottom microphones.** If the tom-tom has two heads, adding a second microphone underneath can add dimension and depth if done correctly. Again, the bottom microphone channel would be switched out-of-phase at the console. Although this adds to the amount of inputs, microphones and time, it can create a solid tom-tom sound.

• **Half your inputs.** Normally, both the top and bottom tom-tom microphone channels are bussed to a single track, then recorded. If certain frequencies need to be added or pulled on the top microphone, chances are those same changes are needed on the bottom, so why not combine them, then process the two microphones as one?

Plug the top microphone into the assigned input at the microphone input panel, and plug the bottom microphone into the same input at another input panel. For example, the top microphone is plugged into input 12, the bottom microphone would be connected to input 12 at a second wall panel in the room. The cable on the bottom microphone would be an 'opposite polarity' cable. This can load down the two microphones, possibly causing low-frequency loss. Use microphones that don't require phantom power.

Save cables and inputs by making a set of 'two-to-one' XLR cables where two 3-foot long cables each have a female XLR connector at one end. One of the female ends has the polarity reversed. This means the connections to pins 2 and 3 are switched. At the other end, these two cables combine and connect to a single male XLR jack.

Mark the reverse polarity connector with red nail polish to show which female connector is reversed. At least that's what you can say you're doing when they catch you buying red nail polish.

• **What's your angle.** Play with the angle and placement of the microphone to hear the best placement with the least leakage.

• **Lost in space.** Pull the microphones back some to capture resonance from the tom-toms that may be lost with close miking. The farther away they are, the more the rest of the drums affect the sound, picking up more of the bulk of the drum, rather than the initial hit.

• **Try this.** Place one microphone between two rack tom-toms and set the polar pattern in a figure-8 position. Of course, check with the drummer that no microphone or stand is in his way.

• **Isolate your low tom.** To get a larger sounding floor tom-tom sound, place foam pads under the feet of the drum. The tom-tom won't lose as much low resonance through the floor.

• **A ring for tom.** To lower the ringing in the toms, toss a handful of cotton balls inside the toms. Ringing decreases depending on how many balls are tossed in. Even properly tuned toms can ring out.

• **Reduce the low tom-tom rattle.** Hang the drummer's stick bag off of the side of the floor tom to reduce rattle.

Miking the cymbals/overhead

• **Don't take the overhead microphones lightly.** The drum sound lies in the overhead microphones. Get a great sound in the overheads, then accentuate this good sound with well-defined kick and snare – key elements of a good drum sound.

• **Condense to the music.** Condenser microphones work well as overheads because they preserve the natural crispness of the sound. Placing them a few feet above the drums will result in smooth response and a warm blend of all the drums nicely meshing.

As the microphones are pulled back, they will reveal more of the room sound. A properly designed room will enhance your drum sound. A poorly designed room will make the best of drum kits sound average. Use the song, proximity to other instruments, and ability of the player to determine the best placement for the overheads.

• **Center the microphones over the snare.** For a centered, focused sound, place two main overhead microphones the same distance from the snare drum. Commonly, two microphones in a stereo pattern are ample for overheads.

• **Height of the mike.** Set both overhead microphones to the same height. From the console, use your eye to match the height of both microphones with a horizontal plane in the studio, such as where the wall and ceiling meet.

• **Do you want to close mike the cymbals?** Dynamic microphones work well when close miking cymbals, due to the combination of high transients and proximity effect. Because close miking cymbals has the potential for maximum overload, pad the microphones, use a peak limiter, and set record levels low. Close miking the cymbals will pick up more low end, but pulling the microphones back and up results in more overall drums being recorded. Other than for close miking, dynamic microphones are not recommended for cymbals or overheads.

If the player is a real basher, pull the cymbal microphones back so less of the splash will come through. Or add a strip of duct tape across the cymbals for damping.

• **You're hat got blown off.** Depending on your needs, either a condenser or dynamic can work on a high hat. But sometimes a high hat microphone is either not necessary or there just aren't enough microphones. Place the snare microphone in a position to pick up more of the high hat, or aim an overhead microphone more towards it.

• **Start with a cardioid pattern.** Drum microphones with a cardioid pattern will pick up more focused areas than microphones with an omni-directional pattern. Big stereo tom-tom fills might lose their stereo effect under omni-directional microphones.

• **Pick up your hat.** To minimize leakage, aim the high hat microphone away from the snare drum. Take time to find the sweet spot where you pick up lots of high hat with minimal leakage from the other drums and cymbals.

• **No rush.** Aim the high hat microphone so it is away from any rush of air that occurs when the high hat opens and closes. A narrower polar pattern may be needed to ease the breeze. Or aim the microphone towards, but not aimed directly at, the center of the bell. This will result in a heavier sound.

• **Use a windscreen on the high hat microphone.** A strategically placed windscreen can minimize any breezes coming from the rush of air into the high hat microphone.

• **Clean cymbals sound better.** If cymbals are dead sounding, clean them with a standard copper or metal cleaner. This returns some if the original sheen, aural and visual.

• **Thin cymbals.** Thicker cymbals are louder than thin ones, so if cymbals are washing over all of your drum tracks, try thinner cymbals.

• **Trashy hat sound.** Some songs may need the trashiest of high hat sounds. For serious trash, ask the drummer to bring in an old cracked high hat. Don't destroy a perfectly good cymbal to do this in case the sound doesn't work with the song. A trashy high hat microphone sound may need a microphone pad.

Miking the room ambiance

• **Condenser microphones usually work best for ambiance.** Dynamic microphones tend to have a midrange bump, so they might not be the best choice. To get a large, ambient drum sound, place a pair of matching microphones in corners of the room, spaced evenly from the drums. Place each a couple of feet from each corner, aimed into the corners to record more reflection and less direct sound. For more freedom while mixing, record ambiance microphones on their own tracks.

• **Set boundaries.** Set two boundary microphones on the floor or on the walls at opposite sides of the studio. These can work well as ambiance microphones, as long as other instruments do not leak into them. I have heard of engineers forcing drummers to tape these microphones to the front of their shirts.

• **Go beyond your accepted recording space.** Place a microphone in an adjacent room, or an elevator shaft, or a loading bay for a natural reverb. Use different microphone spacing and combinations to properly place instruments within the rest of the tracks.

• **Added thud.** Put a microphone in another room just to catch that low-end thump that permeates everywhere. Not really the ambiance, but the low-end rumble. You know that dull thud you hear when your neighbor plays his stereo too loud? Add to the mix for more bottom.

• **Headphones for mike.** Place a set of headphones on a microphone stand across the room. Or hang a set of headphones around the drummer's neck to record the drums from his perspective. Because headphones are transducers, like microphones, they can be plugged into microphone inputs and the stereo signal will be transmitted. Use the proper adapters, and turn off the phantom power.

• **Place the microphones low and in front of the kit.** When aimed upward, these microphones can record an ambient drum sound with ample bottom and minimal cymbal splash.

One microphone setups

Sometimes there is no choice but to use one microphone to record the drums. The fewer the microphones used, the deader the room should be. Figure 3.3 shows three methods to record drums with one microphone. Try:

(a) From above. A microphone placed above the drums has lots of cymbals and snare, but little kick.

(b) In front. Placing the microphone in front results in lots of kick drum and little snare.

(c) In the middle of the room. A distant microphone in the room aimed at the drums is certainly not a tight sound, but more of an ambient sound. Maybe have someone walk around the room with the microphone while the player plays, and listen for the best spot.

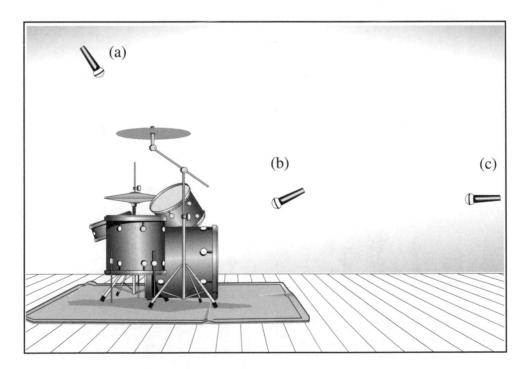

Figure 3.3 *One microphone drums setups*

Two microphone setups

Two microphones open more options for a good drum sound. Whenever you combine two or more microphones, check for phase problems. Figure 3.4 shows three methods to record drums using two microphones.

(a) X/Y overhead. Microphones placed in an X/Y pattern over the drummer record the tom-toms, snare, and cymbals in a nice stereo spread. This setup loses some kick drum.

(b) In front and wide apart. Placed a couple of feet in front of and above the drums, this setup captures the full kit with no real direct sounds. Good but ambient, especially in a live room.

(c) One for the kick and one for the overhead. Determine how close you want the second microphone to the snare drum. Maybe close miking the snare works best, or maybe more of an overhead microphone to record the rest of the drums works best.

 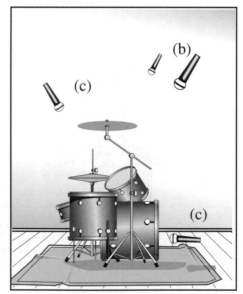

Figure 3.4 Two microphone drums setups

Three microphone setups

Great for jazz, as the overheads pick up exactly what the player is trying to translate – often lost when the drums are individually miked. Figure 3.5 shows three methods to record drums with three microphones.

(a) Kick, and a stereo pair over the kit. This is a standard, and very effective three microphone setup. Many great recordings have been made with this setup. Choose a stereo microphone setup, and listen to what works best in your situation. This setup works well for projects on a limited budget. Fewer microphones mean less time, fewer tracks and fewer distractions. Just take the time needed to place the microphones right.

(b) Kick, snare, and overhead. This setup will help you hear what happens to drums sounds when you raise and lower certain frequencies. This is not a stereo setup as there is only one overhead microphone.

(c) Kick, overhead and distant. Exactly where the third microphone is placed is up to the engineer. As the number of microphones increases, so do your options. Set them up and listen.

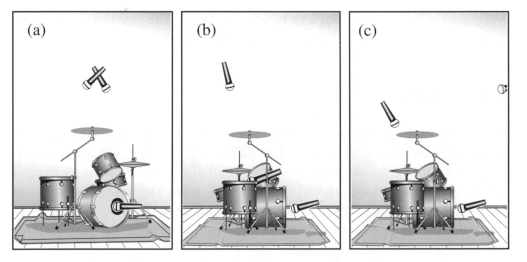

Figure 3.5 *Three microphone drums setups*

• **Use four microphones on the drums.** Commonly, this would be the same as the three microphone setup in Figure 3.5(b) with the addition of another overhead microphone to record the drums in stereo.

• **My aim is true.** Position the overhead microphones so they aren't aimed directly at a tom-tom or at a cymbal. This will avoid one drum or cymbal sounding too loud in one of the overhead microphones.

Or just aim one of the overheads to pick up more of the high hat.

• **Use five or more microphones on the drums.** Close miking the drums is more common today than placing three or four microphones around the kit. To get a great full sounding kit, place one microphone in the kick drum, one on the snare, maybe one on the high hat, and two well-placed overheads. Use the overheads to get spacing and depth between the toms and cymbals. Placed correctly, they will pick up the toms, cymbals and high hat, plus the snare evenly. More microphones mean more time.

• **Don't be fooled.** Great sounds can be gotten with few microphones. Just because you don't have a microphone on each drum and cymbal does not mean the drums will sound second-rate. The fewer microphones you have, the longer it takes to place them exactly right, but a few strategically placed microphones can make a good song sound fantastic.

• **'Sorry, I was in the studio, how much do I owe?'** The best time to go in the studio to check the drum microphones is right when the pizza guy arrives.

Click Track

• **Click track going to the drummer.** A click track is normally sent through the drummer's headphones to help him maintain proper tempo throughout the song. With today's wide array of sounds available, ask the drummer exactly what sound he wants the click to be. Some may want a snare, some may want a cowbell. Maybe he wants the click to be in quarter notes or half notes. Work this out with the player beforehand to make him as comfortable as possible.

• **Don't force players to use a click track.** Some players never use a click when rehearsing or playing live, so they might not do their best in the studio if they are forced to use one. Some players prefer a song's natural ebb and flow to emerge. Perhaps they want to speed up a bit in the chorus, or slow down a bit during a quiet part. A click is not always a necessity.

• **Use a different sound for the off-beat.** Sometimes snare and bass drum mask the click, and he can't distinguish exactly where it is. Sending an off-beat click allows him to hear exactly what the beat is.

• **Some drummers like the click track loud.** Start by turning the click down to a non-lethal level, then raise the level as the drummer asks.

• **Phasing out.** If the click track is leaking into some of your tracks, pan the click to the same place in the mix as the track that contains the click leakage, then switch the polarity on the click track channel. Slowly raise the gain of the click track into the mix until it cancels the offending click.

Click track leakage should be avoided in the first place.

Electronic/Computer Drums

Whether a stand-alone drum machine or DAW-based drums, virtually all studios have some sort of electronically sampled drums. Some engineers use them as timing reference, some use them to trigger real drum sounds, some combine live drums with drum loops. Some producers rely totally on drum machines, while others run at the very sight of one. Whatever works for your situation is the best way.

• **Record live high hat or percussion.** If recording real drums is not possible, at least record a live high hat with programmed drums. Live percussion will help a lifeless electronic drum track because no drummer is exact or perfect, and these slight deviations in timing are what give music its groove.

• **Use individual returns.** Recording every drum to an individual track is great if you have a plethora of tracks. If you are recording all tracks to stereo, return the separate outputs from the drum machine into individual tracks on the console, and buss them over to the two stereo tracks. This allows you to process each drum sound individually so you can equalize it to work with the rest of the instruments. This also allows you to separate the sends to any effects unit so all instruments aren't going to the same reverb.

Return, for example, three different snare drum sounds to the console, then buss the three channels to a single track. Record the snare drum track while changing levels and combinations of the three tracks throughout the song, just to keep all hits slightly different.

Electronic drum sounds commonly bring a lot of low frequencies to the table. Individual returns allow you to equalize individual tracks to make room for other instruments.

• **Punch the drummer.** To give a drum machine more life, ride the different tracks as you record them, maybe push the kick a bit in the chorus, or punch the high hat into a verse, much like a real drummer would do.

• **Use different drum fills.** No two drum fills or tom rolls should be the same unless they are an actual written part. For example, the tom-tom roll into the second chorus of a song should be different, and stronger than the tom-tom roll into the first chorus.

• **Route the electronic drums through a set of speakers in the studio.** Set up two microphones in the studio and aim the microphones into the corners of the room, not toward the speakers. You want ambiance, not direct signal. Add this live stereo track in with the originals.

• **Distortion.** Run the electronic drums through a set of loud distorted speakers in the studio, with close microphones aimed toward the speaker cone. Bring this live sound in just under the electronic drums.

• **Free samples.** There is a plethora of sounds and samples on the Internet. Search out the ones that will be appropriate for your sessions.

• **Get real.** Some engineers place live drums in the room, then send the drum machine signal through the speakers just to record the actual drum overtones and rattles. But if you have real drums, why are you using a drum machine?

• **Get inside the drums.** Many machines process their own drum effects. Access the internal workings and lower the level of the effects processing. Recording dry drums on separate tracks leaves all mix options available.

• **Shy away from hard panning when recording the drum machine tracks in stereo.** Think of an actual drum kit where the cymbal and hat tracks all leak into each other. Drums, especially cymbals, panned hard left and hard right sound unnatural.

• **Don't record the drums.** Low on tracks? Only record the control track that drives the drum machine. This allows you to try different sounds and patterns before committing. Because the drum machine slaves to the control track code every time, even in the middle of a song, there is never any question regarding sync.

• **Bored to run.** Running a repeating drum pattern is boring. Program some changes, such as off-beats, tom-tom fills, appropriate crash or ride cymbals, and all the things that a real drummer would do. Work on parts and beats that will fit within the song.

• **Tom fills in on the drums.** Drummers have two hands, two feet and one mouth – usually big. Program a drum track as if they are being played by a real drummer. For example, don't program a snare drum hit during a tom-tom fill.

• **Alter MIDI.** MIDI is available to control all aspects of the digital drum machine, including velocity and dynamics, program change, etc. MIDI is beyond the scope of this book. There is ample literature available elsewhere.

Chapter Four

THE ELECTRIC GUITAR SETUP

The best guitar sound starts at the source. The guitar (including bass guitar) must be properly aligned and intonated, with all knobs, switches and cables checked and in working order. Most important, it must be in tune. The rest is up to the player. The finest instrument can sound wretched in the wrong hands. You want the listener to get a lump in their throat, not their stomach.

• **Re-cord your guitar.** New properly stretched strings will always get the richest results.

• **For more meat, use thicker strings.** Lighter strings are not as meaty as thicker strings. Thick strings result in a heavier sound, and can be more difficult to play.

• **Bright and oily.** Add a few drops of oil to a tissue and oil the strings to keep string noise at a minimum.

Amplifier Placement

• **Talk to the player.** Before doing any setup of amplifiers or instruments, ask what will make him comfortable when he plays, then set up that way. He may prefer to be isolated with the amplifiers, in the studio, or even with you in the control room. The best placement for the player is where the player wants to be.

Check how loud it is going to be. Some players may play quiet enough to just leave the amplifiers in the room with the drums. If so, then maybe aim the guitar amplifier away from the drums.

• **Use the best amplifier.** If applicable, set up all available amplifiers, and try all combinations. Use an amplifier whose power rating is higher than the speaker. Better the speakers distort before the amplifiers.

• **Place the amplifier as far from the control room as possible.** During basics, the amplifier is often isolated from the rest of the instruments. Unless the control room is isolated from the studio, low frequencies will go through walls, floors, and everywhere.

If you can hear or feel the guitar amplifier through the floor or walls, you won't get a true indication of how it really sounds. If you hear additional lows, you might pull the bass frequencies at the console. When you listen back the sound will be bass shy. Better to record a pass, then listen back to get a true analysis of the sounds before continuing.

• **Space for the bass.** Place the bass amplifier in the biggest room possible. Because bass frequencies are physically longer, and they need space to build. A large dead room works great for the bass as the room echo and ambiance are minimized, and the size allows the low frequencies to build.

• **Try different places.** Place the amplifier at different places in the room, such as a foot away from a wall, and aiming toward the wall. Maybe lean the amplifier over so it rests on a wall and place a microphone under it. Maybe set up baffles around the amplifier to control the lower frequencies or to change the room response. As with the drums, throw a carpet under the amplifier to absorb some reflections.

• **The place for the bass.** Keep a bass amplifier out of corners and away from walls, as the wall increases bass reflections into the microphone.

• **Aim the amplifier into a reverberant space.** Placing a guitar amplifier in a reverberant room, such as a bathroom or kitchen is great for tight natural reverb, but a better result might come from placing the amplifier outside the space, and aiming it in toward the live space. As with the vocals, when the amplifier is placed in the room and you have a close dynamic microphone and a far microphone, lots of these live reflections are picked up by both microphones. When the amplifier is placed outside of the room and aimed in, the ambiance will sound great, but the close microphone will not have all the ambiance. The ambiance track will have more depth than if the amplifier was right in the room.

• **Rising high.** Figure 4.1(a) shows that when the microphone is placed too close to the floor, errant reflections are introduced. Figure 4.1(b) shows the amplifier is raised off the floor to keep reflections from bouncing back into the microphones. Amplifiers generate a lot of vibrations, so unless the setup is very stable, there is a chance of it slowly vibrating off the riser. I've noted that sessions go smoother when amplifiers don't come crashing to the floor.

(a)   (b)

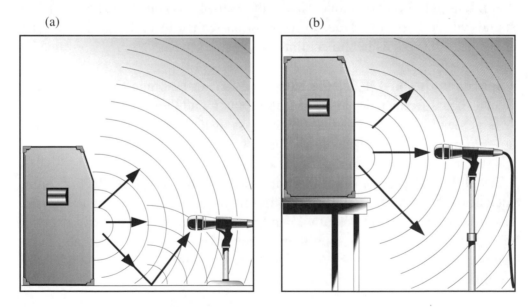

Figure 4.1 *Amplifier on riser*

• **Turn-ups and leaks.** Isolate amplifiers during basics not only to keep the loud amplifier out of the rest of the microphones, but to prevent leakage into the microphone placed in front of the amplifier. Fixing a guitar track that has leakage on it can be tricky, as the repairs won't have the rest of the players in the background.

• **Leave the amplifiers in the bottom part of the road case.** This raises the amplifier off the floor, isolating it. Lock the wheels on the road case or you may have a roving amplifier.

• **Place two amplifiers together.** When splitting a guitar signal into separate amplifiers/speaker cabinets, either totally isolate each speaker cabinet, or place the cabinets beside each other. Placing the cabinets next to each other minimizes delay in leakage. For example, two cabinets in a room are placed next to each other, and both are close miked. Signal, or leakage from cabinet A naturally spills into cabinet B's microphone. Placing cabinet A across the room means the leakage takes longer to reach cabinet B's microphone. These lengthy delays will muddy the signal.

• **Barring grille.** Grille cloths can buzz and metal grilles can rattle. Record amplifiers with nothing between the microphones and the speaker cone.

• **Baffle the amplifiers.** Due to leakage, an amplifier may need to be totally baffled, but no so much that the amplifier overheats. As well, don't isolate it so much that the musician can't get at it. Let him get his sound, then baffle off the amplifier. Occasionally, it's the record producer that's totally baffled.

Connections In and Out

• **Plug in the guitars before turning on the amplifiers.** If amplified, the plug-in transient can blow the amplifier fuses.

• **Keep it short.** Again, keep all 1/4" cables as short as possible. Standard unbalanced cables connected to the high-impedance input of an amplifier should be no longer than sixteen feet, as they lose high-frequency signal with increased length.

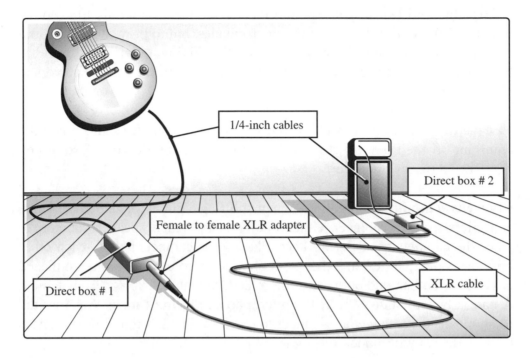

Figure 4.2 *Back-to-back direct boxes*

• **Connect back-to-back direct boxes.** To increase a 1/4" cable length with minimal signal loss, set up two direct boxes with an XLR microphone cable between them. Connect a 'female to female' adapter to one end of the XLR cable. Figure 4.2 shows a short 1/4" unbalanced high-impedance cable goes from the guitar into the first direct box, where the signal switches to balanced low impedance and runs through an XLR cable to a direct box on the other end. Then the signal is swapped back to high impedance, and another short 1/4" unbalanced cable goes from the direct box into the input of the amplifier.

Loud Amplifiers

• **Turn it down.** Many guitarists turn their amplifiers all the way up, then play with the knobs to get the best sounds. Maybe turn the amplifiers down to get a full clean sound, then turn the amplifiers up, adding a natural distortion to your full clean sound.

• **Turn it up.** Some say a good ripping guitar tone can only come from a loud amplifier. Tubes being pushed naturally compress the signal and add a warm presence to a sound. Different tubes can vastly change an amplifier sound. Learning about different tube amplifiers and what they do can aid the engineer in deciding on what works best for what situation. Spend the time getting the right sound at the amplifier, not in the control room.

• **Do not touch.** Ask the player before changing the settings on a musical instrument or amplifier. Note the settings to return to the original sound.

• **Low and behold.** Some bass players add lots of lows when they are playing live, and we love that, but excess low end may not be necessary in the studio. A better sound might come from recording a full round bass, not just the low frequencies. Come mix time, the engineer can fine tune exactly which low frequencies will work best.

• **Distortion-free bass.** Unlike a guitar amplifier, where some distortion may be wanted, a clean and solid bass guitar sound cuts through and sounds best. This means that the bass amplifier might not need to be unbelievably loud to get the job done.

Of course, overload and distortion to the Nth degree is fun too.

• **Use the fuses recommended by the manufacturer of the gear.** Anything less, and you may continually blow fuses. Anything more, and you may blow the amplifier, or worse, cause a fire. If the amplifier is in an isolated booth, this might not be noticed immediately.

• **Ask the player if the output level on the instrument is all the way up.** The cleanest signal is the loudest signal available from the output of the instrument. But some players use the volume control as part of their sound.

• **Virtually a virtuoso.** 'Virtual amplifiers' are plug-ins or stand-alone units that have completely changeable parameters and settings to simulate every major guitar amplifier on the market.

Record the clean output from a direct box to a separate track. At mix time, if you need more 'body' from the guitar sound, run this direct track through a virtual amplifier, and experiment until you have the perfect sound.

Preparation

• **Listen to all the direct boxes available to you.** Direct boxes are like microphones – they differ widely in frequency response. One that works great on a bass may sound sluggish on a guitar. Some amplifiers today have built-in direct-output jacks, eliminating the need for an external direct box. As with most gear, the cheap ones sound cheap, especially in the low frequencies.

• **Try some of the many effects pedals on the market today.** They can give an average guitar sound proper placement. They have modern sounds, and are great to rent for a few weeks while recording. Many all-in-one effects boxes today work as a direct box, as well as a harmonizer, delay, tuner, etc. The signal comes out of the output much like the money comes out of your wallet.

• **Just plug it in.** Sometimes just plug the electric guitar directly into the pre-amp of the console, turn them all the way up to distort the inputs and there's your sound. A high-impedance cable directly into a low-impedance input will overload the pickups of the guitar, changing (some say degrading) the tone. Word is that the famous Beatles 'Revolution' guitar sound was overloaded console pre-amps.

• **Using lots of pedals means various changes in gain.** Go through each pedal and set the levels with as much gain as possible and as little noise as possible without overload. One wrong setting in the signal chain can overload an input.

• **Use new batteries.** Sometimes the effect is quieter with new batteries rather than standard AC power. Write the date on a piece of tape and stick it on the battery. When you are changing batteries, you know how old each one is.

• **Run your signal into a portable boom box that has line inputs.** The built in el-cheapo compressors can really distort a sound. Maybe overload the inputs, or change some equalization settings, then record the output with a cheap microphone for a dirtier sound.

• **Creatively limited.** Of course use compression and limiting to lower peaks and raise overall levels, but also sometimes use it as a creative device to shape a sound.

Miking the Amplifiers

• **Use a microphone that can handle it.** The sound pressure associated with close miking, coupled with the inherent low frequencies of a shredding guitar sound, can damage the sensitive microphone capsule in a condenser or ribbon microphone. A dynamic microphone is sturdy enough to take a lot of level without folding, and has a presence peak that can bring out some of the grit on a guitar sound. A condenser, if it can handle the pressure, can result in a warmer sound. Newer condenser microphones can handle a lot of pressure, but a dynamic is still the sturdier choice.

Avoid overload at the microphone. If necessary, move the microphone, insert pads into the signal or lower the source level. Overloaded signal is unusable and unrepairable unless that overloaded sound is the goal. Digital overload is always unpleasant. Ideally, with the right microphone, record the signal with no microphone distortion and no additional gain. The sound recorded will be cleaner with less hiss and noise.

• **Match the microphone to the speaker.** The different speakers within an amplifier can have different responses, so if the speaker has a lot of bottom, use a microphone with the opposite characteristic – one that is bass shy.

• **Pad early.** Pad as close to the source as possible to minimize overload before the signal reaches the console. Some microphones house removeable pads that you install between the capsule and the pre-amp. Better to pad at the microphone than at the console.

• **The best sound is the sound that works.** There is, of course, a wide range of methods used to set up amplifiers for recording. Suggestions include:

(a) One microphone close. Figure 4.3 shows a common starting point. Place a dynamic microphone a few inches from the cone. A close microphone has a boomier low end due to proximity effect. Pulling the microphone back can ease this. Turning the microphone away from the speaker results in a duller sound. If just a bit more ambiance is needed, try changing the polar pattern to omni-directional. A good spot tends to be right where the cone and the dome meet.

(b) One microphone semi-close. A clean guitar sounds great with the microphone placed one to two feet away and aimed at the cabinet. The proximity effect is defeated, and a certain warmth comes across when the microphones are not so close to the speaker. To capture more room sound than original signal, position the microphone off-axis from the amplifier.

(c) One microphone distant. Use a condenser or ribbon microphone due to the full, dynamic sound. The room itself can be as much of the sound as the instrument, but don't let the milieu of the room overshadow the direct amplifier sound. An ambient sound can easily turn cloudy and disappear in a busy mix.

(d) Two microphones – close and close. Figure 4.4 shows one microphone aimed straight at the edge of the middle cone and another microphone

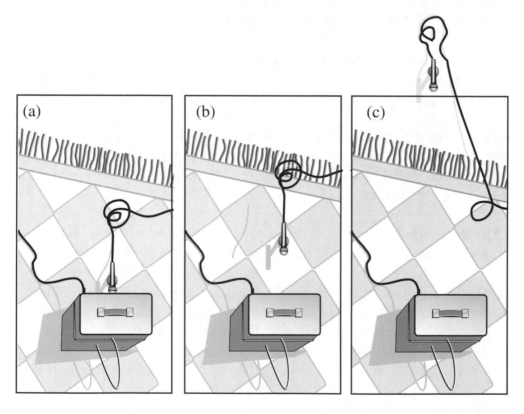

Figure 4.3 *Close, semi-close and distant*

angled away at the outer cone. One picks up more of the crunch, while the other picks up more of the warmth. Sometimes the choice is one or the other, not both. When combining two close microphones, don't expect it to sound twice as good.

(e) Two microphones in stereo pattern. Use this setup for recording either an amplifier in stereo or to combine the two microphones to a single track. Note that some amplifier makers purposely switch speaker polarity to get a more varied sound.

(f) Two microphones – front and back. A second microphone aimed at the back of a speaker cone might add needed tone without introducing equalization. Of course the amplifier must have an open back. This setup has the back microphone aimed somewhat toward the front microphone, so switch the polarity on the back microphone or risk losing signal due to phase cancellation.

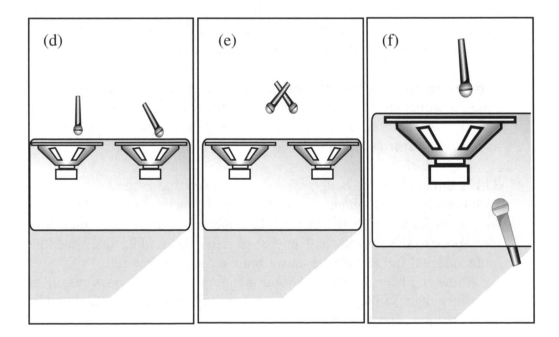

Figure 4.4 *Two microphone amplifier setups*

• **It's a setup.** Additional setups might include:

A direct box. Sometimes there is no choice but to use just a direct box. Direct boxes are great, but there is a certain magic in the air between an amplifier and a microphone. A direct box can enhance flaws that an amplifier in a room may gloss over. String noise, squeaks and clicks come through more on a clean direct track than through the amplifier. But these so-called flaws (that are normally only heard in solo mode) are what gives a track life.

One microphone and a direct box. A close microphone with a direct box is a common method to record an electric instrument, and the standard method to record bass guitar. Record the signal from the direct box and the microphone on separate tracks for more control over the best combination of good tone and tight low end. Check for phase issues between the microphone and the direct signal.

Two microphones – close and far. The close microphone adds definition, clarity and focus, and the far microphone contributes midrange, warmth, depth and placement. A common setup is where the closer microphone is a dynamic and the distant microphone is a condenser.

Two microphones – stereo room. Depending on the characteristics of the room, a proper stereo setup can result in a wide ambient sound. Great for placing a sound at a specific point in the mix, but easily lost in the haze as more instruments are introduced.

Three microphones. Though excessive, some engineers get a great sound with one close microphone, one semi-close, then a third one farther off in the room. Done correctly, they capture the full crunch from the close microphone, the warmth and midrange from the middle one, and the ambiance and full low frequencies from the room microphone.

Some engineers like to record the actual pick hitting the strings on an electric guitar, so they set up an additional close microphone aimed at where the pick hits the strings.

Three microphones and a direct box – holy cow. Going for three or more inputs opens a plethora of options, and these large setups tend to take a lot of studio time. Sometimes too many options can be a bad thing.

• **Consider the polar pattern when setting up to record an amplifier.** A cardioid pattern, rather than an omni-directional pattern on a microphone placed in front of a guitar amplifier might keep the rest of the instruments from leaking into the track.

• **It's time to move the capsule if you dare.** Once the microphones are roughly placed in front of the amplifiers and the player has set the settings on the amplifier, move the microphone around to find the best spot for the music being recorded. To do this:

(1) Have the player turn the amplifier way down.

(2) Turn the microphone up in the headphones.

(3) Have the player play so you can hear more signal in the headphones than from the output of the amplifier.

(4) Hold the microphone in front of the amplifier and move it around until you locate the best spot. Don't worry, you'll hear it.

(5) Place the stand and connect it to the microphone without moving the placement of the microphone – easy to say, maybe not so easy to do.

(6) Ask the player to return his amplifier settings to the original levels.

• **You need some help.** If you have someone helping you, all the better. Ask the assistant to:

(1) Wear earplugs and a set of unplugged headphones. Those high volumes can be damaging to your ears.

(2) Have the player play while the assistant moves the microphone around. Tell him to stop moving the microphone when the player stops playing.

(3) Listen to the microphone in the control room. As the assistant moves the microphone around in front of the amplifier, listen for where it sounds best.

(4) When the assistant locates the optimum spot, you indicate to the player to stop playing.

(5) That's the assistant's cue to stop moving the microphone. He then sets the microphone in that position and tightens the microphone stand, leaving the microphone in the same spot.

• **The final countdown.** If you know you are going to send a guitar to an effect, note that some effects sound edgy in the upper midrange and highs. A guitar recorded with a little warmer sound may counteract that harshness.

• **The wrong microphone.** If the amplifier needs drastic processing, chances are you've got the wrong microphone. Or, an excuse I have heard, 'It's the right microphone, it's just on the wrong amplifier.'

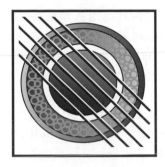

Chapter Five

THE ACOUSTIC INSTRUMENT SETUP

Stringed instruments get their sound from plucking, bowing or striking a stretched string or set of strings. The strings themselves don't generate much power, the power comes from the resonances of the body of the instrument. The size of the instrument, along with the length, thickness and tension of the strings determine the string's pitch and the guitar's harmonic structure.

Naturally a properly tuned rich sounding acoustic guitar will sound better than the old cracked beach guitar. The intonation, neck and frets must be up to professional standards with no buzzes, fret noise or other distractions.

Acoustic Preparation

• **That same G string is getting pretty dull.** Replace old guitar strings to add crispness and clarity to any acoustic guitar sound.

• **Tension in the studio.** Preserve the tension of a stringed instrument by changing strings one at a time. Removing all the strings at once totally eases the tension on the neck – not the guitar's natural state. To allow for proper stretching, change strings well in advance of the session.

• **Start with a good player.** It is difficult to record a player who isn't ready to be recorded. Better a great player with not-so-good sounding equipment than the other way around.

• **Get intimate with the guitar player.** In some situations, do as you would on a vocal recording. Dim the lights, set up a table and music stand to create a more intimate atmosphere. Make the effort to pamper the player a bit.

Placing the Player

• **Lessen lesson.** For an overdub, place the player in the middle of the room, aimed toward the control room. In bigger studios, place a couple of baffles around the back and sides of the player, and perhaps throw a rug on the floor to lessen the room ambiance in the close microphone.

• **How live do you want the sound?** Using an overly dry absorbent space may suck up some of the luster of a guitar, especially if the guitar is not close miked. A natural sounding space with a small degree of inherent liveness, such as a hardwood floor, works wells to record a full sounding track.

For more variation come mix time, some engineers prefer to record in totally dead spaces, eliminating the natural reverb of a room.

• **Hz rentals.** If the instrument is not up to par, consider renting a quality instrument. A rental is a couple of days. A recording is forever.

• **It reflects badly on you.** There are few circumstances when recording in a small, really live room produces the best tracks. Recording an acoustic guitar, or any instrument, in highly reflective places such as the bathroom will always generate lots of short, tight reflections. Maybe wait. If needed, this slapback effect can always be added later.

• **Yo-yo strings.** When recording strings such as violins, placement in the room is critical. Microphones are often placed a few feet away from the instrument to pick up all the harmonics and warmth. It's best to know the room and understand where frequency bumps are. For example, don't place the cellos in a part of the room that has accentuated low frequencies.

Placing the Microphones

• **Start with a condenser.** Condenser microphones work well on all stringed instruments due to the wide frequency range and smooth pickup characteristics. Many dynamic microphones won't have the appropriate high-end capabilities, plus they may introduce certain unwanted frequency boosts.

Sometimes a dynamic microphone is perfect for the situation. Or maybe a ribbon microphone – whatever the situation calls for.

• **Take the tube.** If available, try a tube microphone. A tube microphone has a certain warmth and records acoustics instruments very well. But many of today's small personal studios simply cannot afford tube microphones.

• **Try a large diaphragm microphone** to bring out the inherent low end in any deeper instrument, such as a cello or a stand-up bass.

• **The pattern depends on the environment.** An omni-directional pattern picks up the complete sound field, hopefully making the guitar sound a bit more live. Plus no proximity effect.

The cardioid pattern mostly captures what is placed directly in front of it, and might work better in a poorly designed space.

• **Get down and listen.** Once the player is set up and ready, determine the best placement for the microphone. Because every acoustic instrument and player are different, get down on your knees and put your finger in one ear (yours, not the player's) and listen. As he plays, move around to locate the spot with the richest sound. Place the microphone there.

• **Who wants punch?** Look down the so-called barrel (if applicable) of the microphone and aim it to where the guitar pick (if used) meets the strings.

• **Use your ear as a guide.** If the sound is too loud for your ear, consider a pad. Recording, for example, a light acoustic guitar you probably won't need a pad. Recording a trumpet, you might.

• **Don't aim a close microphone directly in front of the sound hole,** but off-axis a bit. As the player plays, the sound hole resonates at a low frequency. This will create a woof in an improperly placed microphone. Aim the cardioid microphone capsule more toward the neck and less toward the sound hole so any 'woof' is naturally minimized by the microphone's polar pattern. If you must aim the microphone into the sound hole, place a windscreen in front.

• **Check your microphone levels.** Place the microphone so it picks up all the strings at the same level. Placed incorrectly, the recording might have just the high strings with not enough impact. If the guitar is there to simply add sparkle, you may want less impact. Maybe lower the microphone, or pull it away to capture more of the higher strings and harmonics. Since the idea is to get the best sounds without a lot of processing, record the instrument appropriately for how it will be placed in the final mix.

• **Aim the microphone away from the player's mouth.** Any unwanted breathing noises may spoil an otherwise quiet passage.

• **Move it.** While you listen in the control room, have someone slowly move the microphone around in front of the sound hole to get the best placement.

• **What works for this may not work for that.** Final placement depends on the style of music. A jazz acoustic guitar would not be miked the way a pop acoustic guitar would be. A country picking guitar would use a different placement than a love song. Slow songs would be miked differently than fast songs. All situations are unique. Figure 5.1 shows:

(a) Close miking. Place the microphone one foot or closer for a 'harder' sound, with an emphasis the upper midrange. Proximity effect and sound hole resonance might be a concern. As well, the player must stay still for the sound not to change. Moving the microphone farther away from the sound hole should give the player a bit of freedom to move without compromising the guitar sound.

(b) Not so close miking. Move the microphone farther away from the guitar in an acoustically suitable room for a rounder warmer sound. The microphone will pick up all the parts of the guitar that are missed with close miking. A single microphone placed from twelve to eighteen inches away and aimed toward the instrument is a reliable starting point. But a guitar with too much room sound gets lost in a busy mix.

(c) Pickup microphone. A smaller microphone attached to the guitar ensures a uniform sound whenever the player moves around, as the microphone stays the same distance from the instrument at all times. This setup works well in a live room because the microphone picks up the acoustic guitar and little else.

(a)   (b)   (c)

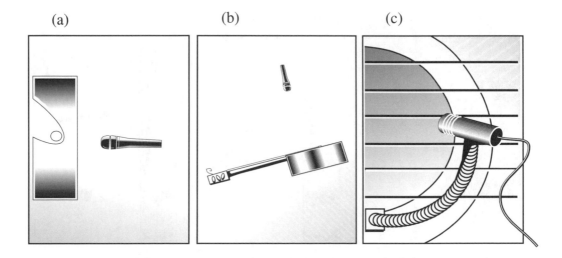

Figure 5.1 One microphone acoustic setups

• **Use two microphones on the acoustic.** Common placement for two microphone setups on an acoustic guitar include:

(a) Close/far. Figure 5.2(a) shows a second, distant microphone is added to increase ambiance and placement. Due to the sound's natural delay in reaching the second microphone, check the phasing between microphones.

(b) Stereo. When using two or more microphones on an acoustic guitar, the temptation is always to make a real stereo sound. When two microphones are placed to capture a stereo spread, a hole in the middle becomes more apparent as the distance between microphones grows.

(c) Neck/sound hole. Set up and connect an additional microphone to record the string noise, then combine it with the other microphone. One microphone captures the higher frequencies, and the other microphone gets the lower frequencies. Adjust for proper highs and lows with the level faders rather than with equalization. Either use matching microphones for a uniform sound, or two different styles to capture different tonal nuances of different areas of the guitar.

(a)   (b)   (c)

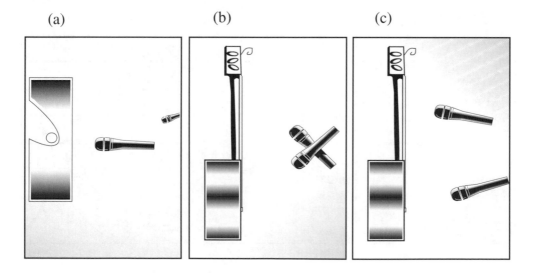

Figure 5.2 *Two microphone acoustic setups*

• **Pickup the microphone.** If there is a pickup in the acoustic guitar, run the signal through an amplifier, and then place a microphone in front of the amplifier. Combine the microphone on the instrument and the microphone on the amplifier at the console, changing it to more of an electric sound than an acoustic one.

• **Recording a guitar and vocal.** When the player wants to sing along but you don't want the leakage, try a pickup on the acoustic guitar, and tight polar patterned microphone for the vocals. A close microphone on the guitar works, but there will be leakage issues on the vocal microphone. The vocal sound is more important than the guitar sound.

Some engineers place a microphone stand or Plexiglas plate somewhere between the player's mouth and hands to minimize leakage, but this adds reflections, and results in the player being uncomfortable – the last thing you want.

• **Room size vs. close miking.** In a large room, use two microphones on a five-piece string section. In a smaller room, maybe individually mike them. Placing the microphones closer to each instrument results in less of the room influence, and more of the direct signal.

• **Combine the direct and the miked signal.** Use a microphone to capture all the rich lows, and the direct signal to capture the nice highs. Acoustic pickups are great for live situations, but in the studio, if you have a choice, a well-placed microphone will sound better. Perhaps record the pickup on a separate track, then pan them left and right for a stereo effect. You decide.

• **What, this old thing?** Violins, cellos, violas, etc. can be expensive instruments, and the players are normally very good and very expensive to hire. Come in early to make sure you are totally set up and ready to record when they walk in the door. Better string players are used on jingle sessions, where they are on a strict time schedule for the day.

• **Don't close mike violins.** Violins really only sound full and rich if there is a few feet of air between the instrument and the microphone. A single player might not warrant a stereo microphone setup. To record a two or more violins, place cardioid microphones in an X/Y position quite a few feet above the violin players for a rich, ambient stereo sound. As the string section grows, cellos and violas get miked from the front, not above.

• **Use the riser wiser.** In the early days of pre-amplification, stand-up bass players would use a riser, or raised platform when they performed. Because the instrument was resting on the riser, the riser itself would resonate enchancing the overall low frequencies, and making the bass appear louder.

• **String a guitar differently.** String an acoustic guitar with the higher octave strings from a 12-string, and have the guitarist play along with an original acoustic track. Even better, have two guitarists play the same part together while miking both. This may give more depth to the track. Playing a 12-string gives the 6-string track a chorus-like effect.

• **Cover the sound hole for a deader, dryer, sound.** Of course, this stops the air pressure from getting into or out of the sound hole. The minimal air pressure generated inside the guitar has nowhere to go, so no air pressure means no microphone capsule movement. Compressed air itself makes no sound at all, except when my grandfather is involved.

Piano

• **Good news, there's no wrong way to do it.** The piano is such a versatile instrument, there is really no right or wrong way to record it. Different microphone placements will, of course, result in different sounds. Using a certain setup for one song may work great; using the same setup for another song may not work at all. The wide range of piano styles makes it impossible to say one method of recording is wrong and another method is not.

• **Tuning the studio piano.** Recording engineers don't normally tune the piano. Leave that to a professional.

• **Place the piano before the piano tuner arrives.** Moving the piano after it has been tuned may throw it out of tune.

• **Aim the open lid toward the open studio, not toward a nearby wall.** You don't want reflections bouncing back from the wall into the piano microphones. Move an upright piano well away from the wall.

• **Check, please.** Check all the pedals for squeaks or noises and with the player, eliminate them by oiling the pedals. Check with the proper people before dousing the pedals with oil. Maybe a pillow stuffed under the pedals will work, but the player might object.

• **Back to that leakage thing.** Sometimes the piano must be recorded in the main room, with the rest of the instruments. Eliminate leakage using baffles and packing blankets. If possible, isolate the piano in its own room. Easy to say, but few studios have a dedicated piano booth.

• **Reflect on the day's work.** If the player uses no print music, remove the built-in music stand on the piano, but leave the lid. The reflections from the lid are part of a piano sound.

Placing the Microphones

• **Open lid – insert head.** Once the piano is correctly placed, put your head in the piano and listen to the player play the music to be recorded. Consider the part to be played, the player's ability, the style, the tempo and prominence of the piano part to be played before placing the microphones.

• **What style of microphone?** Microphones vary between large and small diaphragm, each having their own role. Large diaphragm microphones might work well for more classical and jazz or slower tempo songs, while small diaphragm microphones might work better on rock songs with faster tempos. Whatever style of microphone you choose, use the biggest heaviest strongest microphone stands available.

Tube microphones create the warmest sound and tend to have more of a pleasing harmonic distortion than solid-state microphones. Condenser microphones work well to capture the wide range of frequencies produced by the piano. Dynamic microphones may capture the piano's lower frequencies best. Boundary microphones will minimize leakage when taped to the underside of the lid, and the lid is closed or on the short stick, covered with a moving blanket.

• **Close or distant microphones?** Because the sound board – where most of the sound is generated – has a large surface area, really close miking may not capture the full spectrum of the instrument. Close miking the sound board will add a brightness that cuts through other instruments and features more 'hammer hitting the string' sound. Mid-placement microphones add tone, while distant microphones contribute more room sound. As the distance between the microphone and the sound board increases, the more influence the surrounding area has. In a properly designed acoustic space, distant miking works very well to capture a full rich piano sound.

Maybe close miking might work for a pop session and a more ambient sound may work for a jazz or blues session. Often, the style of music is the final determination for placing the microphones, except perhaps when the piano is baffled off and part of a full session.

• **Use shockmounts on piano microphones.** Heavy instruments with an abundance of low end can make the floor and the microphone stand rumble. Without shockmounts, the microphones pick up this low rumble.

• **Use one microphone on the piano.** Reasons for recording the piano in mono include not enough microphones or inputs at the console. Perhaps the piano is just a support instrument among many and would get lost in stereo. Perhaps the mix plan has the piano panned over to one side, so a stereo recording isn't needed. Maybe there is just not enough tracks. As a starting point for setting up one microphone on the piano, Figure 5.3 shows how one microphone might be:

(a) A microphone angled in from the side is aimed sideways toward the high and low strings using a figure-8 pattern.

(b) A microphone placed across the middle of the sound board perhaps with an omni-directional pattern.

(c) A boundary microphone is taped to the inside lid of the piano. Use caution with duct tape on the piano, as it may ruin the finish. Maybe tape it to the floor or the wall. Maybe.

(d) A microphone is placed a few feet away from the opening of the piano, just enough to capture the attack of the keys, but not so much as to render the sound overly ambient.

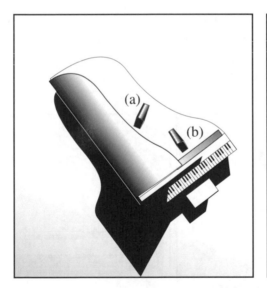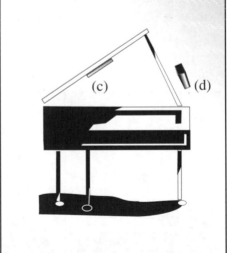

Figure 5.3 One microphone piano setups

• **Use two microphones on the piano.** Commonly, two microphones do a fine job of recording the grand piano, capturing the complete range of notes across the board. Figures 5.4 and 5.5 show how different setups might include:

(a) A stereo pattern. Try a well-placed stereo X/Y pattern about a foot above the sound board, using one microphone to capture the highs, and the other to capture the lows. This setup results in a bright in-your-face sound. As well watch the spacing so those all important middle notes aren't lost in the hole between the microphones.

(b) High and low. Use one microphone aimed toward the higher strings and another aimed toward the bottom, or low end of the piano. Some pianos, such as upright, or baby grand pianos may lack sufficient low end. A large diaphragm microphone on the lower strings just might capture more of the lower frequencies.

(c) Across the sound board. Aim two microphones straight down at a point about a foot above the sound board at 1/3 and 2/3 of the way across.

(a)   (b)   (c)

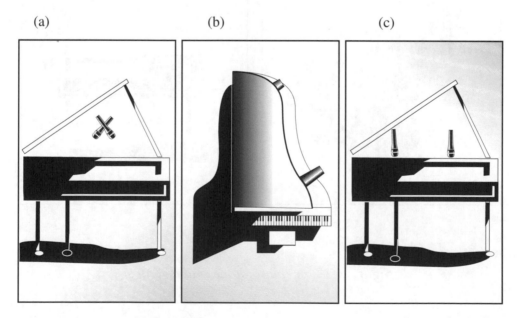

Figure 5.4 Two microphone piano setups

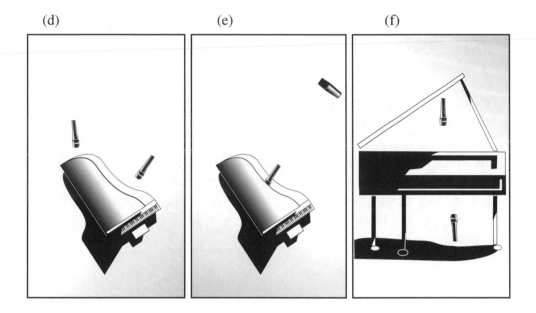

Figure 5.5 *Additional two microphone piano setups*

(d) Both microphones a few feet away. Set them raised high enough to aim into the piano, maybe or maybe not set in a stereo pattern.

(e) One microphone close and one far. Not a stereo recording, but sometimes the piano might not need to be recorded in stereo. Record too many stereo tracks in a song, and they might all melt into one blob of cloudiness.

(f) One microphone aimed at the strings and the other aimed up from underneath. Again, because the microphones are aimed at each other, switch the polarity on the bottom microphone.

• **Raise the piano lid.** When close miking, raise the lid, if applicable, as high as you can to minimize the reflections from the lid back into the microphones. Don't raise the lid too high if a boundary microphone is taped to the inside of the lid.

• **Use three microphones on the piano.** As two microphones hold a stereo pattern over the sound board, a third microphone might capture low end, especially on the grandest of grand pianos. Or use a third microphone as a room enhancement to any the above stereo setups.

• **Listen to the final sound in mono.** If some frequencies disappear, you have phase issues between microphones.

• **Talk to the hand.** Record some piano and let the player hear it just for the sounds. This is good for all instruments of course, but because the piano can have such a wide variety of sounds, the player, especially a soloist, can be picky. The player is often happy to voice an opinion. Consider all suggestions or comments. A happy player plays better.

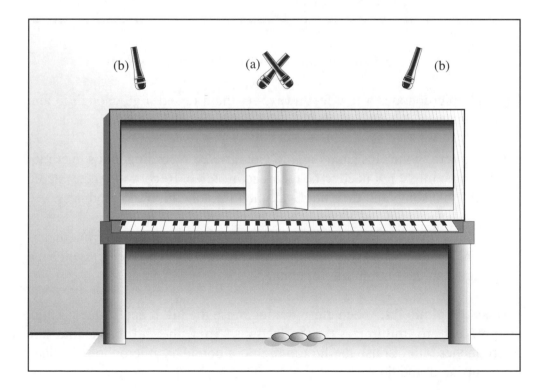

Figure 5.6 *Upright piano*

• **Upright is not as grand.** Recording an upright piano can be a challenge, as the strings are hard to access, and the piano is usually up against a wall. Moving it away from the wall may ease reflections into the microphones, but it may change the amount of perceived bass.

The low end that makes a grand, or baby grand piano sound so rich is the nice long, low piano strings. The string length is limited on an upright piano. If possible, remove the top of the piano. Figure 5.6(a) shows an X/Y position and Figure 5.6(b) shows two spaced microphones over the top of the piano. You could also lower a microphone into the piano, but due to proximity to the strings, some notes would be much louder.

Keyboards

• **Prepare.** Ensure that all the prep work is done beforehand, with everything programmed and prepared. Hours can be wasted by inexperienced people learning how to operate equipment on the client's time.

Unless you or the programmer/player are totally familiar with the instrument keep the manuals handy. Whoever is sequencing and programming the computers should know what they are doing.

• **Confirm that all sequences are correct before committing.** Ideally, you want to press play, and have the sequence run through totally correct. As well, some situations require the sequencer to lock up or sequence to a pre-recorded time code track.

• **Place a limiter across the outputs of an unfamiliar keyboard,** especially if you are experimenting with sounds. One blast of a piercing sound can ruin your speakers, and it won't help your ears much either.

• **Go direct.** Modern keyboards have either an XLR or TRS phone-jack balanced output, so it can be plugged into the line inputs of the console. Older units may need a direct box, and should be plugged into the microphone inputs to raise the signal to a usable input level.

Horns

• **Use a live room to record the horns.** Nothing sounds better than good horns in a live room – but not too live or else standing waves and unwanted ringy overtones may pop up. Toss rugs across the floor and set up baffles to minimize any errant frequencies.

• **Place the horn section to get a nice live ambient feel.** Omni-directional microphones are great for distant miking things like horns and strings, as the natural reverb of a room comes through.

   But placing the horn section too close to a wall or window may result in too much refection being recorded. When the horn players are lined up and aimed at the control room the sound may bounce off the window and back into the microphones.

• **Set up the horn section the same as they would on stage.** Place all the players in a row and set up a pair of microphones a few feet away. This setup will capture the warmth of the horns and eliminate some of the shrill sound associated with close miking. They will tell you how they prefer to be set up.

• **Recording the ribbon section.** Ribbon microphones are popular on brass sections because of their great high-end transient response and smooth warm lows. Just don't place them too close to the horns. Ribbon microphones often can't handle the serious pressure generated. Dynamic microphones work well as they capture the sizzle generated without overload.

• **Let's try my pad.** Horns tend to range, in volume, from loud to really loud with lots of transients so microphone pads may be needed. Listen before pressing the record button, then listen back for distortion.

• **Keep it a foot away.** For a less clicky and warmer overall sound, keep the microphone at least a foot away from the bell of the saxophone. The air between the instrument and the microphone contains life. Placed too close the wind can really 'pop' the microphone.

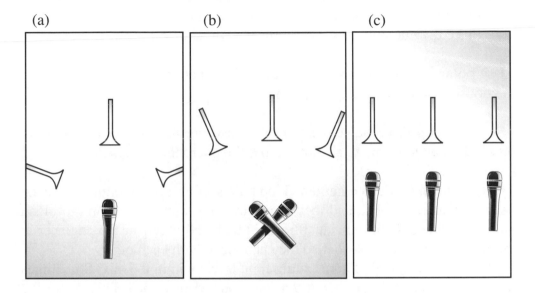

Figure 5.7 Horn placements

• **How may microphones are needed to record horns?** Different setups for a horn section may depend on the amount of microphones you choose to use.

(a) Use one microphone. Figure 5.7 shows how one microphone may have all players standing around it and playing.

(b) Use two microphones. Two microphones may be placed in a stereo pattern, or in tandem with all players is a row.

(c) Use three microphones. A three microphone setup may allow each horn to have its own microphone. In this case, follow the three-to-one rule of microphone placement.

• **Let's radiate those transients.** Because horns radiate a lot of high transients from the bell, placing the microphones directly in front will result in a hard shrill sound. Aiming the microphone at an off-axis angle, such as 45 degrees, or moving the microphone a few feet away from the bell, should result in a warmer pleasing sound, as well as minimize any key click sounds associated with some horns.

• **Lowest of the low.** For some wind or brass instruments, such as clarinet, flute or saxophone, sounds can come from the whole instrument, not just the bell. Place the microphone so it records the whole instrument.

• **Use polar patterns to your advantage.** Polar patterns may play a key role in choosing and placing a microphone.

• **Set up a pop filter.** Some wind or brass instruments may need a pop filter or windscreen to decrease air flow across the microphone capsule.

• **Record horn sections together.** Good horn sections, like groups of singers, need to hear each other, and lock in to how they want their sound to blend with each other. With individual tracks, this blend is left to the engineer at mixing. Record them together so they can blend with each other.

Listen to the blend of horns at different levels within the mix. Sometimes horns can sound full and big when loud, but when turned down, only the trumpets are heard.

Proper microphone choice and placement should allow the whole sound to come through at a low level within the mix. With improper microphone choice and placement, only the shrill parts tend to come through the mix when the horns are set at a lower level.

Chapter Six

THE VOCALS SETUP

When setting up for the vocals, the most important factors are the singer's comfort and confidence. Preparing a vocal station is the key. The singer should be able to walk in, put on her favorite set of headphones, and sing the vocals with everything exactly right. Set out whatever will be needed to keep her relaxed and comfortable, perhaps:

– A carpet on the floor and a chair or stool.

– Some candles or incense.

– Posters of their influences.

– A small table with some water, tea or spirits.

– A box of tissues and a trash bin.

- Dim the lights and set the temperature of the room.

- A copy of the lyrics with a pencil for any last minute changes.

- A lighted music stand to see the lyrics.

- Perhaps a daring photo of yourself.

• **Set up a vocal station for the duration of the project.** Or maybe have it ready, but placed to one side. When the time for vocals hits, the assistant can quickly grab it and move into place. On days when the singer gets the urge to belt one out, everything will be ready to go.

• **Use an open space with baffles.** To get a tight dry sound, place baffles a few feet behind and beside the singer. The surrounding environment should be live enough to create a natural ambient sound, yet not overly reflective.

Some engineers prefer a totally dead sound, others go the other way, using a larger area for the vocal to give it natural depth and placement. If the singer is in a natural sounding room, a hint of ambiance can give a vocal track more life – especially background vocals.

• **Avoid tiny spaces for vocals.** A claustrophobic environment can do more harm than good. The singer's comfort will take precedence over microphone technique. As well, placing the singer in a small booth may color the sound. A good microphone in a small booth might pick up the lower midrange room reflections, especially with male singers as their voices tend to be lower.

• **Don't use the bathroom.** Put a singer in a small absorbent situation, rather than a reflective situation, such as a washroom. As with acoustic instruments, once those natural echoes are recorded on the vocal track they cannot be removed. That effect can always be recreated if the vocal is recorded in a more natural sounding environment.

• **Mike's in the ladies room.** Another reason not to place a microphone in the washroom is the privacy issue. People expect privacy. Plus, some rather dodgy deeds have been known to occur in recording studio washrooms.

Placing the Microphone

• **There is no one microphone that works best for everyone.** For a jazz session, you might want a warm full sounding microphone. For a rap session, you may want an edgier, brighter microphone. A female singer might get a different microphone than a male singer.

• **Does the sound sound sound?** Maybe a player with a degree of studio experience will request certain microphones on their vocal or instrument. If you disagree and use a different microphone, guaranteed the player will not like the sound. Keep him happy and use his microphone, even if you feel the sound is compromised.

• **The singers hide.** Some singers do not want to be seen when they sing – even by the engineer. Ask him if he would be more comfortable behind a screen, or facing away from the control room, or maybe in total darkness. It's his choice.

• **Choose your pattern.** If recording one singer, a cardioid pattern may work best. Two singers opposite each other may need a figure-8 pattern. With four singers around one microphone, an omni-directional pattern may work best.
   If the singer likes to jump around a lot in the studio, maybe try an omni-directional pattern.

• **I'm shocked. Shocked!** A microphone installed on a proper shockmount eliminates much of the room rumble. When no shockmount is available for the vocal microphone, use a roll-off. Roll-offs are internal microphone filters, and can color the sound.

• **Hang 'em high.** Depending on the microphone, try hanging it from a boom-stand so the capsule dangles, not sits. This can also make room for the singer to see the music stand, plus he won't be tempted to hang the headphones on the stand when the vocal track is done.

• **Patience is a virtue.** Take a few extra minutes to set up a few microphones and placements to determine which works best with the song and the singer's particular style. Figure 6.1 shows six common methods of placing the vocal microphones:

(a) Six to eight inches in front. This setup tends to be a great starting point, as the distance adds a warmth that may be lost with close miking. It can mean less compression is needed if the singer knows what he is doing and can control his dynamics.

(b) Quite close. For a very intimate vocal, place a singer's microphone a few inches away from the mouth to record all the little nuances of the vocal. This works when the vocal track is dynamically consistent, and the singer doesn't suddenly yell into the microphone. Both condenser and dynamic microphones record well when close miked. Perhaps set up a windscreen.

(c) Over a foot away in front. Placing the microphone farther away, beyond twelve inches, captures the 'air' that can get lost from being too close. Condenser microphones might work better than dynamic microphones. This setup works if the singer moves around a lot while recording due to the distance between singer and microphone.

(d) Perpendicular to the nose. Place the microphone so the front, or pickup surface is positioned directly parallel with the top of the singer's nose and aimed at or around the singer's mouth so no nasal tone will be registered. The off-axis placement keeps sibilance from hitting the capsule square on. A windscreen is suggested, but may not always be necessary.

(e) Perpendicular to the nose, over a foot away. Works well in small absorbent spaces and in very dynamic situations, where vocals go from soft to loud. As well, the singer can read the lyrics if the microphone is not directly in the way.

(f) Positioned underneath and aimed up at the mouth. Some singers like to lean into the microphone when they sing. Place the microphone a touch lower than the mouth, so the singer's head can bend down a bit. As well, this is a natural way to capture resonance from the chest and body cavity.

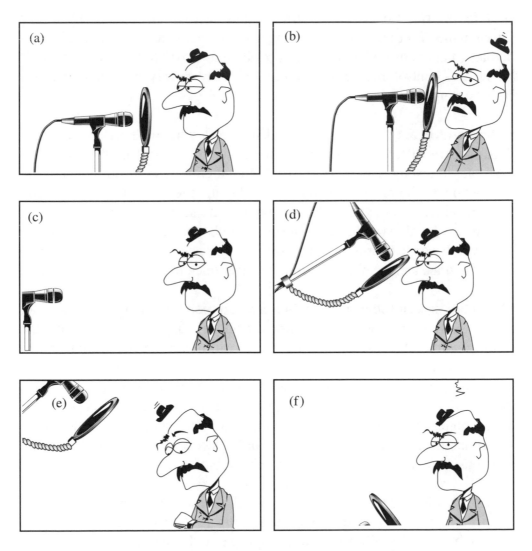

Figure 6.1 *Vocal microphone placements*

• **Background vocals.** Use as few microphones as possible with background vocals. Try a single omni-directional microphone or a pair of microphones for a group of singers. When they can see and hear each other, they will create a proper blend of harmonies between themselves. Listen for who is louder or softer, and move each singer closer to or farther from the microphone until the best balance is reached.

• **Strip on the floor.** Run a strip of tape across the carpet where the microphone is set up and ask the singer not to cross that line as he sings. If he has great control of his dynamics, place him closer to the microphone. If he has less control, maybe place him a foot or two away, and remember this when setting the limiter.

• **Oral and hearty.** On those unusual instances where the singer gets very dynamic, try different microphones for different sections of the song. For example, in a song with quiet verses and loud choruses, set up two adjacent microphones where one microphone has the input level set lower. Have him sing the song, then record each microphone onto separate tracks.

One track will have the verses just right with the choruses too loud and the other will have the chorus just right, with the verses too quiet. Comp the verses from one track and the choruses from the other track to a separate master track.

• **Who has screen the wind?** A windscreen, or pop filter stops the flow of air across the microphone's capsule. Some engineers argue that this colors the sound so rather than use a pop filter, they position the capsule away from the singer's mouth.

To keep a singer from getting too close to the microphone, maybe attach the pop filter to its own stand, and place it at the proper distance from the microphone. This creates a barrier to allow for sufficient space between the singer and the microphone. The best recordings come from the singer staying at a consistent distance from the microphone.

• **Wash it, will ya?** Occasionally, replace the mesh across the pop filter and wash or replace the windscreens, especially if they are being used on a regular basis. With different singers spitting and breathing into it, it may become a little rank.

• **Keep pops out of the studio.** As a last resort, attach a pencil in front of the microphone so it splits the pressure and diverts it away from the membrane inside the capsule.

• **Take time to find the best pre-amp for the vocal.** Just as microphones have different sounds, so do pre-amps, equalizers and compressors. Due to the nature of physics, every device contributes a tiny amount of distortion. Inexpensive equipment such as cheap pre-amps contain inexpensive circuitry, which introduces distortion earlier.

• **Listen to the microphones through a set of headphones.** Because digital recording is so clean, the quietest air conditioner or the smallest toe tap will be clearly picked up. Listening with headphones will expose any errant sounds coming through the microphones.

• **Give the singer tight fitting closed-back headphones.** If a singer wearing open-back headphones leans in too close to the microphone, the resulting feedback can do serious damage to the singer's ears, and to your perceived ability to control feedback.

• **Reflections.** Place a towel (or something similar) on the music stand to absorb the reflections so the signal does not reflect back into the microphone and color the sound.

• **Use the force.** All singers are different. Some have fine technique, and know how to work a microphone. Others use brute force. Try different placements out, and find the spot that is most comfortable with the singer, yet works for you sonically. When the singer is happy with everything, the session goes smoother and everyone benefits.

Chapter Seven

THE CONTROL ROOM SETUP

In the studio room, opening wall panels, moving baffles, laying out carpets, and hanging blankets are common methods of altering the acoustic space for your specific needs. In the control room, the opposite is applied. The room response is set and predicted. In a properly treated room, the reflections are minimized so sound reaching the engineer is not colored by the influence of the room characteristics. The environment remains consistent to accurately monitor what is being recorded and replayed.

Control Room Analysis

• **Tune your room.** A spectrum analyzer shows an accurate readout of a room's frequency response. To analyze a control room, a technician with a trained ear places a microphone about where the engineer's ears will be, then plays pink noise through the speakers at various volume levels. The spectrum analyzer displays a readout of the room's frequency response as modified by the reflections.

The technician might insert a graphic equalizer between the amplifier and the speakers to 'flatten' the room. He might pull a resonant frequency of the control room in order to decrease any unwanted influence.

Not unlike using an electronic tuner to tune a musical instrument, the spectrum analyzer is just an aid. Ultimately, the technician's trained ear will determine what sounds correct.

• **Place all equipment in its final positions before tuning the room.** Once the room is tuned, moving machines and racks of outboard around may change that tuning.

• **Listen, man.** Beginners may not be sure of what sounds 'right', so when the technician is tuning your room, have him explain what he is doing. Listen to and note what frequencies are boosted or pulled, and why.

• **Understand the room's limitations.** If you can't physically change the control room or the speaker equalization, at least understand the deficiencies so you can aurally compensate. For example, if you know the room has an untrue low-end response, you might want to push the low frequencies a little more than you normally would. This takes well-trained ears, and usually comes only with years of experience.

• **What is a bass trap?** A bass trap is a dissipative device placed or built at the back, sides or corners of a room to absorb a room's resonant frequencies, and therefore its harmonic intervals. Something as simple as another room behind the control room would be considered a bass trap. Once a control room has its resonant frequencies controlled, and has the correct mid- and high-frequency absorption to get uniform reverb time versus frequency, it is referred to as a neutral room.

Control Room Preparation

• **Keep the room clean, organized and stocked.** A clean work environment shows that you are serious about your job. A messy space makes people more anxious, and if the place is a total wreck, clients think your tracks are the same way. No matter how much or how little clients are paying, they expect quality.

Stock the room with colored felt pens, duct tape, energy bars, band aids, earplugs, aspirins, vitamins, guitar strings, a drum key, pencils, pens, track sheets, labels, console tape, speaker fuses (especially when recording loud drums) and all the peripheral items needed to keep the session going. Place all computer documents in the same file, and keep everything well labeled.

• **Get the lead out.** Place a notepad and a couple of sharpened pencils on the console. While recording, jot down the counter number on any mistakes or questionable bits, and check them when appropriate.

• **Put an air purifier in the control room.** A clean smelling control room is so much nicer to work in. Portable air purifiers can change a room from rank and dank to clean and pristine.

• **Who's on first?** Mute the console and headphone amplifiers before powering up or down. Power surges can blow fuses, speakers and headphones. Power amplifiers are the last things turned on before the session, and the first things turned off at the end of the day.

• **EQ placement.** Place outboard gear, especially equalizers, where you can adjust them while you are within the stereo spectrum. If they are across the room, you won't be able to hear the changes as you make them.
    Equipment needs plenty of ventilation so everything stays nice and cool.

• **Confirm equipment rental arrangements.** If you sign for the rental, check the equipment for any dents, scratches, or flaws, and note the serial number. Like renting a car, if it comes back with a dent, you are responsible.

• **Check your gear.** Use an oscillator to verify that all recorder and processor inputs and outputs are correctly connected to the console. Listen to check if all the processing equipment is indeed coming through as it is supposed to, with no loading, incorrect patches or unwanted processing at the console. If something is not working, you want to know before the session begins.

• **Open says me.** Turn on the computers and open the proper programs. Check that all inputs and outputs are correct and the system is ready to record.

Digital Recording

Due to low cost and high quality results, all studios today have some kind of digital processing. The function of digital recording is the same as analog. Record in real time, then edit and mix. Our only limitation is the amount of free space on our hard drive, and the amount of simultaneous inputs/outputs that our audio interface or recording software is able to deliver.

• **Place the DAW in the middle of the monitor speakers.** In some studios, the computer workstation has its own set of monitors. In others, the output signal is routed to the main studio monitors. Place the computer monitor screen so you can see it while facing the monitor speakers.

• **Use the highest quality sound card available.** The less expensive cards use cheap A to D converters that do a 'grainy' job to your sounds.

• **Turn that thing off.** Shut off the power before moving the hard drive. The internal discs are always spinning, so moving the drive might corrupt data.

• **Set parameters with the project in mind.** Before recording to hard drive, establish the parameters of use, such as bit depth, sample rate and reference level. The session setup windows must be addressed when a new session is started. Some of these cannot be changed later.

• **Few standards.** The analog world has well-established standards, but the digital world has few. There are many horror stories of the incompatibility of files from one DAW not working on another DAW.

Use the following guidelines to make sure you have a smooth running session. It helps if you know what the final output medium will be. What is the recommended:

> **Sample rate?** For CDs, 44.1 kHz is the required sampling rate, so recording at 44.1 kHz avoids the signal degradation of downsampling.
> For DVD A (Hi definition audio) use 88.2 kHz, 96 kHz, 192 kHz.
> For DVD (Video) use 48 kHz.
> For T.V. use 48 kHz.

Bit depth? 16-bit or 24-bit. All modern playback systems can handle 24-bit. With the low cost of modern hard drives, the small amount of space saved using 16-bit might not be justified. The only time to select 16-bit would be for final CD-audio mastering.

Reference level? In video and film, -20 dBFS peak level corresponds to 0 VU average, or RMS level. In digital audio there are no standards. The most common ones are -16 dBFS or -18 dBFS for the multitrack part of the project. When mixing down to stereo or surround audio -12 dBFS or -14 dBFS = 0 VU.

• **What is a clocking reference?** Almost all professional digital equipment has the ability to have the master clock either derived from the internal clock, or synchronize to a central, or external clock. Some cheaper digital processing equipment lacks a high quality, stable internal clock that generates the sample rate. Using a higher quality external clock will vastly improve the sound of the converters and plug-ins of your digital processor.

• **What are plug-ins?** Digital plug-ins are small apps that process the digital sound file. They represent outboard effects that in an analog world would be part of the studio's outboard rack. Equalizers, compressors, gates, reverbs and delays are 'plugged in' after you have recorded the sounds flat to your hard drive. In the analog world this would be like recording all your sounds flat to the multitrack recorder, then processing the sounds on the monitor panel part of the console.

• **Plug-ins will create a small delay in the sound file.** Every plug-in has a different processor delay (normally stated in samples) called latency. Delay can be from a few samples to many thousands of samples depending on how complex the effect is. This timing error can become noticeable if you add many plug-ins to, for example, the snare and nothing to the kick. What was a tight drum track is now skewed somehow. Slide it in time with the original sound file, then remove the original to be sure that all is still in time. Some DAW software corrects plug-in latency.

• **I've created a master.** Create a master file with everything properly labeled, bussed and assigned. Start a new session at a moment's notice with everything set up just as you like it.

• **If applicable, format enough tapes or hard drives to keep rolling.** There is no reason to run out of tape on a session. Nothing kills a vibe more than 'Sounds great. Wait while I format some more tapes so we can get started.'

• **Use the most expensive digital tapes.** Cheaper tapes shed, are less reliable, and will create machine maintenance issues sooner rather than later.

Speakers

• **What is a loudspeaker?** A loudspeaker is a transducer that changes electric current to acoustic signal. Not unlike a dynamic microphone, a paper diaphragm houses a voice coil that hovers within a magnetic field. The voice coil reacts to the applied current, causing the diaphragm to recreate the equivalent waveforms.

• **What is a studio monitor?** A studio monitor is, at least in most studios, more than one speaker. Because one speaker cannot reproduce the complete audible frequency range, different speakers cover different ranges. Often larger recording studios will have large mounted speaker arrays aimed at the engineer, as well as various nearfield monitors. Figure 7.1 shows a common three-speaker studio monitor with two amplifiers. This includes:

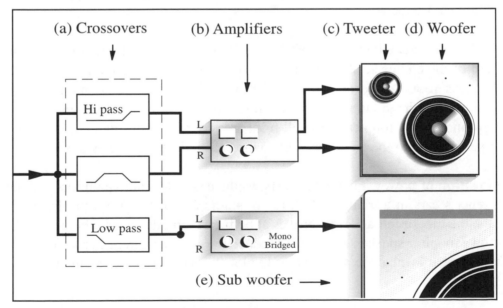

Figure 7.1 *Studio monitor*

(a) Crossovers. A crossover is the network of pass band filters that splits the signal feeding the individual speakers, sending only the high frequencies to the tweeter, and the low frequencies to the woofer. Some crossovers are active, or amplified, to allow the user to set the individual speaker level, the frequency of the transition, and the slope of the crossover. Larger mounted studio monitors may have more than two speakers, each with its own amplification.

(b) Amplifiers. In this example, the first amplifier's left side powers the tweeter, and the amplifier's right side powers the woofer. When a second amplifier is used for the subwoofer, the stereo output is commonly bridged for a louder mono signal.

(c) Tweeter. A small speaker used for high frequencies. The crossover pulls all low and mid-frequencies before reaching the tweeter. Larger mounted speakers sometimes use a 'horn' for a tweeter.

(d) Woofer. A larger speaker used for low to mid-frequencies. Frequencies beyond this speaker's rated range are rolled off.

(e) Subwoofer. A large heavy-duty speaker designed specifically for the very lowest frequencies.

• **What is a nearfield monitor?** Nearfield monitors are placed a few feet in front of the engineer, often on stands just above and behind the meter bridge. Whereas room shape and treatment heavily influence the sound from big mounted studio monitors, nearfield monitoring is influenced far less by the surrounding environment. Still, all front areas of the room should be designed to push sound to the engineer's ears with as little interference as possible.

• **Position of power**. Figure 7.2(a) shows the nearfield studio monitors and the listener's ears in a triangle. With the monitors equidistant from the listener's ears, the mix of left and right arrives at both ears simultaneously. Figure 7.2(b) shows monitor speakers on stands produce fewer console reflections. Woofers being at ear level are less crucial, as the low frequencies are less directional.

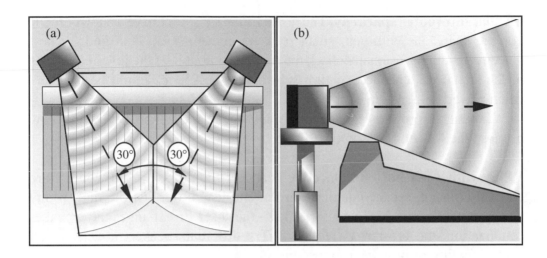

Figure 7.2 *Nearfield monitor placement*

• **Keep your speakers free of debris.** Any items placed on top of your nearfield monitors will eventually vibrate off, landing on the console.

• **Keep speakers out of corners and away from walls.** Close proximity reflections can cause interference with direct signal, creating phantom images. If it is not possible to move the speakers, then deaden the corners to reduce unwanted reflections.

• **Rainbows everywhere.** Keep your speakers away from video monitors. The magnets, if unshielded, mess with the colors on the screen.

• **High ratio hornblower.** Use the best speaker cables available, and add in-line fuses to your speakers – always install the recommended fuse. High peak levels, such as when the drummer hits the kick, will translate into high voltages and can overload the speaker. Hopefully, the fuse will blow before the speaker, saving the speaker. Replacing a fuse is cheaper.

• **Use a powerful monitor amplifier.** Low powered amplifiers will overload sooner, creating unwanted distortion. The speakers should distort well before the amplifiers distort. Many engineers today use nearfield monitors with built-in amplifiers.

• **Add a subwoofer sparingly.** The home listener may not have a subwoofer. If you are using one with your small speakers, move it around to find the best placement, and occasionally check the sounds without it. If the low frequencies in the studio are enhanced with a subwoofer, the sounds might be bass shy when played somewhere else.

Without a properly connected subwoofer, small speakers can blow because the low end is too powerful.

• **Play with your bottom.** Once the nearfield monitors and subwoofers are correctly placed, set the equalization flat and listen to a favorite disc. Many professional nearfield monitors allow the user to change the speaker equalization and crossover frequency. Try different speaker models for an overall feeling of how the room sounds.

When we know that the room and speakers are properly equalized, we can rely on our ears knowing that what we hear is a correct representation of what is being recorded. The room should be designed and tuned so all frequencies are kept under control – no major frequency bumps or dips exist and all outside noises stay out and inside noises stay in.

Console

• **Understand the console.** Of course, the console is the routing station where all incoming signals are sent on their merry way to wherever you choose, such as to the multitrack recorder, the monitors, or the auxiliary cue sends. The console enables you to use equalization, panning, effects and dynamic range as building blocks for the best possible sound.

And remember the terminology, the console has channels and the multitrack recorder has tracks. Commonly, the signal is called the track.

Whether it is a 12-channel console, a 48-channel console, or a DAW, they all share the same principle. The channels process incoming signals such as microphones, direct signals from instruments, returns from the multitrack recorder and returns from effects. Signals are then assigned a location, and finally mixed to a master output, often stereo. Without a total understanding of the console flow, the engineer cannot use it to its maximum potential.

• **What are VU and peak meters?** VU, or volume unit meters average the signal level much like the human ear does. VU meter readings correspond with perceived volume. Peak meters show the peaks, or the maximum signal of transients that do not register on the VU meters. The peak meter readings accurately show the recording levels.

• **Label the console.** Use a scribble strip to clearly label the channels on the console with all relevant information including microphone and line inputs, multitrack returns, cue sends and effects. Use a strip of tape across the length of the console, as it tends not to smudge. A separate strip of tape for each song can be pulled off the console and taped to the wall for future use, saving time when changing from song to song.

• **Don't get loaded down at the bay.** Clear the patch bay of all patches. Improperly grounded external equipment can cause hum. Partly plugged-in patch cords could cause channels to 'short' out, or crackle.

• **Mute the channel.** Mute the channel before making a patch. Dirty patch cords can cause grounding buzz, click or crackles as the patch is made.

• **Check the microphones.** Once the console is set and the inputs are properly routed, have a runner scratch the head of each microphone to check that each one is properly connected. Or have him snap his fingers in front of the microphone. Some engineers prefer finger snaps, especially the jazz engineers.

• **Check the headphones.** Go out to the studio room and listen to all the headphones to make sure the cue sends are connected and checked for each individual player. Each player's headphone set should have ample signal level, with everyone hearing the talkback microphone.

Equalizers

Most recording studios have racks of outboard signal processors including equalizers, compressors, limiters and effects. Equalizers process frequencies. Compressors, limiters, and noise gates process dynamics. Effects such as delays, reverbs, and choruses process time. See Chapter Eleven for more on effects.

• **The trouble with trebles.** Learning how to equalize is a lifelong pursuit for the recording engineer. The more you listen, the more you realize that every sound has a complex waveform structure, and each sound interacts differently with other sounds. Although we cannot hear the very high and low frequencies, they still interact with the frequencies we can hear, and contribute to the overall sound. Figure 7.3 shows the equalization frequency response broken down into:

(a) Very low. 80 Hz and lower contains rumble and mud. Low E on a regular guitar is 82 Hz. Low E on a bass guitar is 41 Hz. Below that is only rumble from trucks or air conditioning.

(b) Low. 80 Hz–350 Hz contains an instrument's trunk and body. Boosting in this area brings out natural fundamental tones.

(c) Low midrange. 350 Hz–2 kHz contains body and meat of a sound. Most instruments have substantial harmonics here.

(d) High midrange. 2 kHz–6 kHz contains placement and definition. High-mid frequencies can bring presence to a track and can bring it to the forefront. When pulled, they can cause the track to blend into the background.

(e) High. 6 kHz and higher contains crispness, clarity and presence. The higher frequencies, from 12 k on up, are known as 'air' for their brilliance and high sheen. Overuse can cause symptoms not unlike an icepick to the forehead.

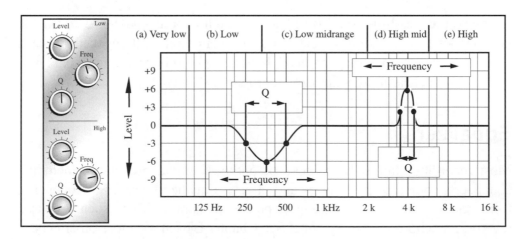

Figure 7.3 Equalizer

• **What is Q?** Q, for quality, defines the width of the bell curve that is affected when boosting or cutting equalization. The bell curve, so named because it is shaped like a bell, contains the frequencies slightly above and below the center frequency. These adjacent frequencies are affected as signal level is raised and lowered. Today's computer plug-ins allow the user to change the bell curve to much steeper values than traditional analog design.

• **Q the music.** The Q value equals the center frequency (in Hz) divided by the bandwidth (in Hz). Bandwidth is measured at -3 dB of the peak of the chosen center frequency. When cutting, bandwidth is measured at -3 dB below the horizontal axis, resulting in thinner Q value.

Figure 7.4 shows a wide Q setting (two octaves), a medium Q (one octave), and a medium thin Q (half an octave).

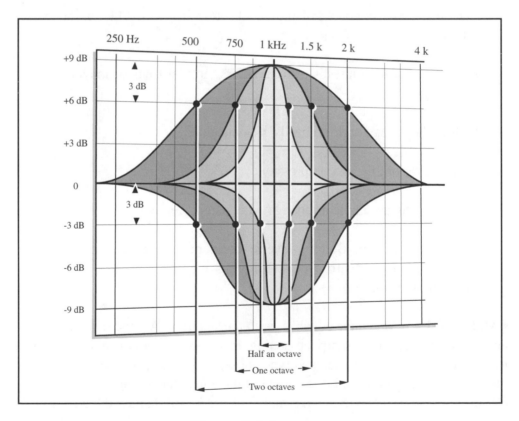

Figure 7.4 *Q settings*

Equalizing

• **Don't twiddle.** Ask yourself 'is equalization necessary?' Go into the studio and listen to the actual instrument, or sound source. As the player plays, listen to what frequencies are apparent in the studio, but missing in the control room speakers. The ideal situation is to get the right sound at the source, then record it with as little processing as possible – not always easy to do. Before reaching for the equalizer, maybe:

– Move the microphone. Higher frequencies are more directional, and sometimes a small placement change can make a world of difference.

– Change parameters on the microphone. Maybe a pad or roll-off could be installed or removed.

– Change the microphone. All the equalization in the world cannot replace using the right microphone.

– Ask the player to adjust his instrument or amplifier tone, or maybe even his playing style.

– Replace the strings or drum heads, if applicable.

– Add, remove or change compression.

– Try something as simple as changing the pan placement.

– Replace the player. Better players sound better, so they are easier to record. A recording is forever, so bringing in a professional session player is not uncommon.

• **Match and set.** Match the equalizer with the situation. A good equalizer is rich and musical, yet razor sharp at honing in on certain frequencies. As with microphones, different brands and types of equalizers can be used for different instruments and situations. For example, if the equalizer is being used to simply roll-off some low frequencies, you might not need a full bandwidth equalizer.

• **Smear campaign.** Adding equalization also adds phase shift, a degradation in the waveform that increases with added equalization. This 'smearing' can cause a track, especially a vocal, to lose clarity in the final mix. Plus adding high end boosts hiss and noise that comes from the microphone, the console, the outboard equipment, and all else along the audio chain.

• **If you must equalize, keep it musical.** Use the musical content of the instrument to determine where to add and pull frequencies. Each musical note's complex waveform is made up of its fundamental tone, plus that tone's harmonics and overtones. The richness of a sound lies within these overtones and other harmonic frequencies, so adding and pulling random frequencies just adds non-musical clutter. If, for example, the song is in the key of A, there will be a lot of buildup around 220 Hz and its harmonics (440 Hz, 880 Hz, etc.). Logically, these harmonics are where the meat of the sound resides. Figure 7.5 will help you find specific musical tones and overtones.

Octave 0		Octave 1		Octave 2		Octave 3		Octave 4	
Note	Freq.	Note	Freq.	Note	Freq.	Note	Freq.	Note	Freq.
C	16	C	32	C	65	C	130	C	261
C#	17	C#	34	C#	69	C#	138	C#	277
D	18	D	36	D	73	D	146	D	293
D#	19	D#	38	D#	77	D#	155	D#	311
E	20	E	41	E	82	E	164	E	329
F	21	F	43	F	87	F	174	F	349
F#	23	F#	46	F#	92	F#	184	F#	369
G	24	G	48	G	97	G	195	G	392
G#	25	G#	51	G#	103	G#	207	G#	415
A	27	A	55	A	110	A	220	A	440
A#	29	A#	58	A#	116	A#	233	A#	466
B	30	B	61	B	123	B	246	B	493

Octave 5		Octave 6		Octave 7		Octave 8		Octave 9	
Note	Freq.	Note	Freq.	Note	Freq.	Note	Freq.	Note	Freq.
C	523	C	1046	C	2093	C	4186	C	8372
C#	554	C#	1108	C#	2217	C#	4434	C#	8869
D	587	D	1174	D	2349	D	4698	D	9397
D#	622	D#	1244	D#	2489	D#	4978	D#	9956
E	659	E	1318	E	2637	E	5274	E	10548
F	698	F	1398	F	2793	F	5587	F	11175
F#	739	F#	1474	F#	2959	F#	5919	F#	11839
G	783	G	1567	G	3135	G	6271	G	12543
G#	830	G#	1661	G#	3322	G#	6644	G#	13289
A	880	A	1760	A	3520	A	7040	A	14080
A#	932	A#	1864	A#	3729	A#	7458	A#	14915
B	987	B	1975	B	3951	B	7902	B	15804

Figure 7.5 *Frequencies to tones*

• **Pull before adding.** Pulling a midrange frequency, then raising the overall level often works better than adding highs and lows. The response is smoother with less risk of overload.

• **Freq. out.** Minimize frequency overlap by pulling offending frequencies that mask more important sounds. Pulling a frequency on one track changes the way other tracks interact at those frequencies. Figure 7.6 shows the frequency response of equalization applied to two guitars.

On guitar (a) around 250 Hz is being pulled with a bump around 4 kHz to give the guitar some presence. On guitar (b) around 250 Hz is brought up, and somewhere around 4 kHz is pulled to let the other track shine through. When all frequencies are accounted for, with no overlap, the result is a tonally balanced sound with both definition and clarity.

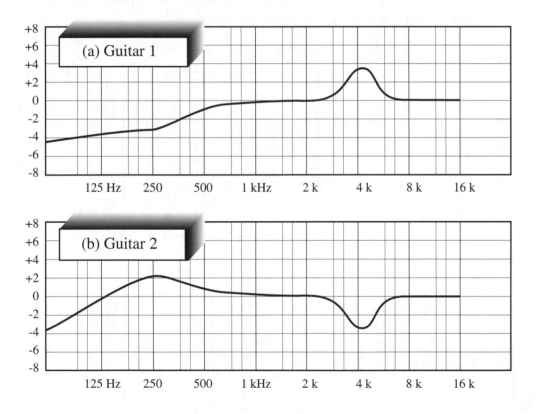

Figure 7.6 *Minimize frequency overlap*

• **Q tips.** Proper Q settings can play a key roll in creating distinction between instruments. Set too wide, it affects frequencies that might not need affecting. Set too thin and it may not get the job done.

• **Thin Q very much.** As you sweep to locate certain emphasized frequency ranges, use the meters to determine where the peaks begin and end. Set your Q according to these points. The meters are only an indication, and should not be used as the final determination.

• **I boosted it, then I cut.** To locate specific frequencies, the tried and true still stands the test of time.

(1) Lower the track level.

(2) Pan the track to center it between the monitor speakers.

(3) Have the player play the part. Listen for the general frequency range you wish to address.

(4) Turn up the level of the equalization, adjust the Q and sweep the equalization until you determine which frequency range needs to be addressed, whether it's muddiness being pulled, or crunch being added.

(5) Find the right spot, then return the level and the Q to a less drastic position. A wider Q pattern can result in a less 'pointy' sound.

• **Use a wide Q when adding equalization so less level is needed.** When pulling equalization, keep the Q setting narrow, so the richness isn't lost by cutting too wide a swath. This may help preserve the tone a little, rather than make the instrument sound lacking.

• **No peaking.** Excessively high equalization levels, such as plus or minus 8 or 10 dB or more rarely sound good, and high peaks are a nightmare for the mastering engineer.

• **Too much equalization can overload.** Because equalizer filters are active circuits, adding level to one frequency adds level to the overall signal. Channels easily overload when there is an abundance of equalization.

• **Find and pull the instrument's resonant frequency.** Some instruments, such as the bass guitar can have pronounced boom at certain frequencies. Sweep the general areas where you feel these resonant frequencies are, and if necessary, pull them a bit.

• **Enhance a frequency.** Boost (or cut) at double (or half) a certain frequency, making the desired change a bit smoother. For example, if you are boosting at 220 Hz add a bit of 440 Hz, and pull off the 110 Hz a bit.

• **Zero must tell.** Find a specific frequency to change by pulling out all the low frequencies. This will help zero in on the midrange frequencies. Once set, bring back the low end.

 • **Watch your bottom.** On budget consoles, the equalization settings may not be sweepable, but fixed. If the kick drum, the snare drum, the tom-toms, the bass, the guitars and the vocals are all boosted at 100 Hz, they will meld together as one indistinct murky blob. Better to use microphone choice and placement to get your low frequencies just right.

• **Don't overequalize.** Think of the painter who wants his painting to be very bright, so he mixes the brightest greens, with the brightest reds, yellows and blues. Individually, they are all very bright. Combined, they turn into a mass of dull brown and gray.

Each instrument must be laid out on the canvas so all colors can be seen and the whole canvas is used – even if some spaces are left white.

• **Overequalize.** There are times when drastic equalization is needed. The trick is to know where and how much. When a feature instrument is not as large as it should be, maybe that's the time to lay it on thick. Maybe.

Compressors

• **What does a compressor do?** Most musical instruments and vocals are so dynamic that a compressor is needed to raise the lowest levels and lower the highest levels. Like an automatic level control, this process raises the apparent loudness of the signal, bringing out the quiet notes that may otherwise be lost. But raising low levels also raises the noise floor. The challenge is recording the hottest signal level with the lowest noise possible. Compressor controls can include:

Input/output. Most compressors have, at the very least, a level control. Raise the input too much and overload the input circuits. With not enough level, noise is introduced when the compressor output signal is raised.

Threshold. The threshold determines at what point the compression begins to take effect. Any incoming signal that exceeds the threshold gets compressed. A higher threshold means less of the signal is affected. Figure 7.7(a) shows a high threshold where minimal signal is affected. Figure 7.7(b) shows a medium threshold, and Figure 7.7(c) shows a low threshold, where most of the signal is affected.

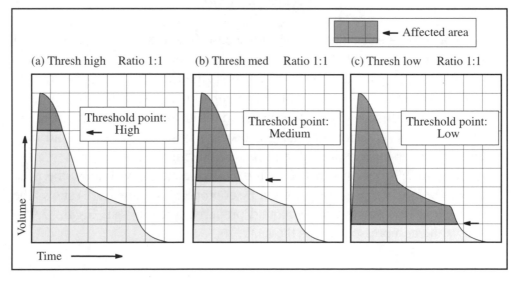

Figure 7.7 *Threshold*

Ratio. The ratio is the dB change in the compressor's input signal compared to the dB change in the compressor's output signal. The envelope in Figure 7.8(a) shows that a 1:1 ratio is unity gain, or no change from input level to output level. Figure 7.8(b) shows a 2:1 ratio. For every 2 dB coming in, only 1 dB change goes out, once the threshold is reached. Figure 7.8(c) shows an 8:1 ratio. When the ratio reaches 8:1 or 10:1 the compressor becomes a limiter. A high ratio combined with a low threshold results in severe change in sound and level.

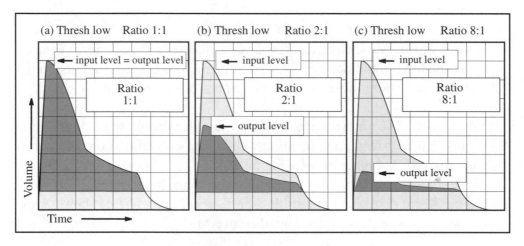

Figure 7.8 Ratio

Attack and release. The attack and release functions define the speed of the compressor action. The attack time is the time the compressor takes to react to a strong signal. The release time is the time the compressor takes to return to unity gain after that strong signal has past. Some instruments take a few milliseconds for the power to build, so their peak point is not instantaneous. A snare may have a faster release time and a piano may have a slower release time, due to its natural ringing out.

Figure 7.9(a) shows that if set too slow, the attack will allow the peak to pass before compression kicks in. If set properly, Figure 7.9(b) shows the attack and release control the sound properly. Set too fast, Figure 7.9(c) shows the compressor misses the peak.

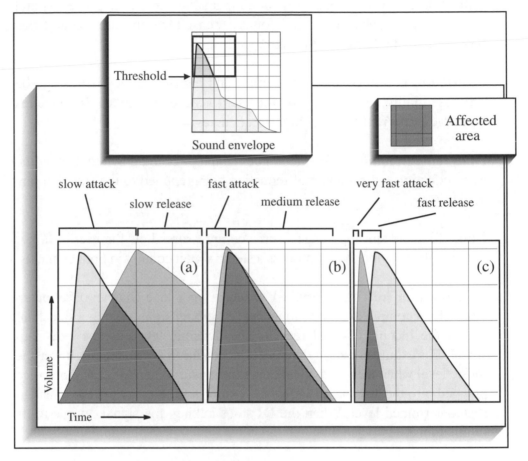

Figure 7.9 *Attack and release*

Additional Processing

As with equalizers, compressors vary in their build and application, giving the user a wide range of options, including:

Soft knee. Also called over-easy, this setting eases the compression in rather an abrupt bump. This gradual transition makes the effect less noticeable.

Hard knee. The compressor has an abrupt transition from unity gain to compression at the threshold point. Hard knee preserves the attack element, or hardness of a sound, where a soft knee would soften the transition.

Split. Some compressors have a circuit that splits the signal, processes and distorts one to enhance the harmonics, then allows the user to add this processed signal back into the signal path.

Multiband compression. Low-, mid-, and high-frequency bands can be processed individually, allowing the user to compress the low frequencies on a track different than the high frequencies.

Gain reduction meter. The meter goes from 0 VU and down to display the gain reduction – the amount of reduction being applied to the signal at any given time.

Link. To avoid shifting images, some compressors allow the user to link it with another compressor to process a stereo signal equally in both channels.

Sidechain, or key in, or duck. Used to 'trigger' one channel off another. Normally a compressor reacts to its own incoming signal. Using a sidechain makes the incoming signal react to the dynamic levels of the sidechain signal. For example, radio DJs use the sidechain to automatically lower the music level when they want to talk over the music. When they talk, the voice plugged into the sidechain triggers the compressor, and lowers the music to a predetermined level. When the DJ stops talking, the signal returns to its original level.

De-esser. A compressor with a frequency-dependent sidechain. A de-esser compresses only high midrange frequencies that contain sibilance, such as 'S' and 'T.' The de-esser is used mainly on the vocal track to reduce sibilance while still retaining the crispness and high end.

• **Different compressors include:**

Peak limiter. A limiter is a compressor with ratios of 8:1 or 10:1 or higher. Used to control the very fastest and loudest of signals, it stops the signals' output from exceeding a certain level, regardless of the input level. The peak limiter works on the peaks, or initial spikes – the loudest initial part of a sound, rather than RMS (average) as other compressors do.

Leveling amplifier. This is a basic compressor with preset medium attack and medium release times, and only input/output level controls, with stepped ratio control.

Vacuum tube/valve. Vintage microphones, amplifiers and limiters may contain vacuum tubes, also called valves. These tubes distort musically, and create a warmth that makes tube equipment a favorite for many engineers.

Compressing

• **Choose with care.** As with microphones, compressors must be chosen for each situation. The right compressor with the correct settings will improve intelligibility and placement, making the track appear louder.

• **If adding equalization place the compressor before the equalizer.** This allows you to fix the equalization settings without having to compensate on the compressor. But maybe the pre-amp in the equalizer is better than the pre-amp in the compressor. This may mean that the equalizer comes first in the chain.

• **If pulling equalization, place the equalizer before the compressor.** Pull unwanted frequencies such as low-mid muddiness so the compressor reacts to the equalized sound, not the pre-equalized sound. Sometimes the equalizer can't handle the incoming peak levels, so the sound needs to be compressed first.

• **Don't overcompress.** Too much compression tightens up the air and can eat all the dynamics of the track and produce a lifeless sound. Some engineers love to use lots of compression as part of their sound.

• **Overcompress.** Let's say you have a great guitar sound, but it needs more attitude for that real in-your-face feel. A low threshold and high ratio, together with proper input and output settings, will 'squash' the sound for a more ripping tone. Too much compression can suck all the dynamics out of a track so the output level will stay the same no matter what the input level is.

Overcompression can enhance the low frequencies, but it steps on other tracks in that frequency range. Once the compressor settings are as you want them, re-examine the equalizer settings.

• **First in line gets set up first.** If you have set up a compressor, then a peak limiter, then an equalizer, start with the settings on the compressor.

• **Set the attack according to the instrument and the music.** A string pad might have a slower attack time than, say, a jarring staccato violin piece. Consider the instrument, the music, and how the music is played when setting attack time.

• **Set the release time to match the tempo of the music.** A faster song would likely have a faster release time. Too long though, and the release will still be recovering when the next signal hits.

A longer release time can smooth an instrument's decay. Too short a release time results in pumping and breathing – the sound of the release returning to unity gain too early.

• **No compression, but no overload.** Add another limiter to totally stop any signal from going beyond a certain point, but without 'compressing' the signal. Set a very high ratio such as 10:1 (the point of limiting) and set the threshold on the highest setting to allow everything to go through unaffected. Play the signal at its loudest, and lower the threshold until you see meter movement. Go back and forth between ratio and threshold until you find the right settings for your situation.

• **Peaking duck.** The limiter output signal level should have the peaks just reaching -1 dB. Set the output level so it doesn't quite reach the overload point, yet captures the maximum signal level with minimal noise. The ideal situation is to set the limiter's input level so that the dB-reduction meter is not moving during the main musical parts. When a loud part hits, the compressor kicks in.

Noise Gates

• **Gate with the program.** A noise gate, also called an expander, works in the opposite way to the compressor. Unless a certain input level is reached, the trigger will not open. As with a compressor, a noise gate includes features such as threshold, attack and release. It may include a 'hold' setting which tells the unit how long to wait before the release, and a 'duck' setting, which activates the sidechain. Noise gates, like compressors, are commonly placed within the chain of a signal track, not across a send, such as a reverb would be. Figure 7.10(a) shows the incoming signal, (b) the gate settings, and (c) the outgoing signal.

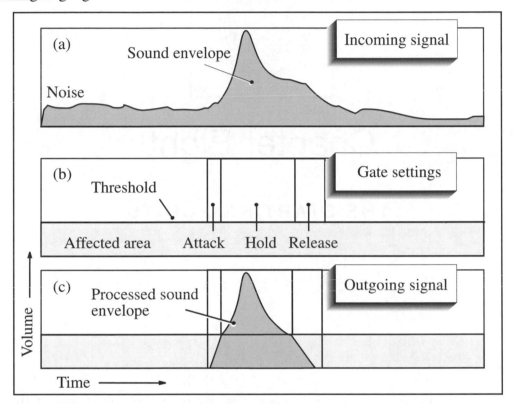

Figure 7.10 *Noise gates*

• **Hum free.** A gate will only eliminate hum when the instrument is not playing. Once the player starts, the gate opens, and the signal, with the offending hum, comes through, loud and clear. Best to eliminate all the noise before pressing the record button.

• **Install a reverse gate on the talkback microphone.** When the players play, the volume in the studio will cause the gate to react, in effect turning off the talkback microphone. When the music stops, the gate opens and activates the talkback microphone. This allows more freedom than always having to reach over and continually turn the talkback microphone on and off.

Chapter Eight

THE STARTING POINTS

Whether you are recording analog or digital, the incoming signals need to be routed, recorded and monitored. Depending on the style of music, the artist, and the tracks available, decide how many tracks will be needed for basics, how many will be needed for overdubs, and how many will be needed for vocals.

Track Layout

Determine which tracks get higher priority. For example, if the primary artist is a drummer, more tracks than normal might be allocated to the drums. If the player is primarily a pianist, record the piano in stereo rather than mono.

These choices are always a challenge, because you must peer into the future to imagine the final outcome. Using the number of available tracks, you need to know:

– How many instruments will eventually be recorded.

– Which instruments get routed to which tracks.

– How many tracks will each player get. Often drums get the most tracks. For example, a 48-track session might use numerous tracks for the drums, giving each tom-tom (maybe both top and bottom) plus the snare drum, cymbals, and ambiance microphones their own track. However, with only 16 tracks total available, you might want to assign drums to say, five tracks – kick, snare, hat, then a stereo mix of tom-toms and cymbals.

• **Lay your tracks out consistently for every session.** When the tracks are always laid out the same, you get used to where everything is. The console will become your instrument to play, like a piano. Rather than think, 'If I want that, I have to do this,' your instincts will take over. As well, you may be able to leave the setups the same from song to song, saving time.

• **Layout on the console.** If you are using a digital console, banks of tracks are layered. Assign stereo tracks evenly, such as not having one half in one bank, and the other half in another bank. As well, you wouldn't want, for example, the effects for the drums returning in a block that has the vocals. Historically, effects have returned on one section of the console. With digital consoles, effects for the drums should stay with the drums, and effects for the vocals should stay with the vocals.

• **Track trick.** Route the vocals to higher channels so they naturally return to the console more in the middle rather than all the way over to one side. Recording the vocal on track one usually means leaning way over to address it. Better to be centered when you dial in those all-important vocal settings.

• **Maximize your choices.** If tracks are available, try recording different sounds of the same instrument. For example, record a bass track direct, a bass track with a microphone through an amplifier, and a bass track going through some effects. Use different processing on different tracks and come up with a suitable mix. Any or all can be erased or combined later.

• **Hey, my stereo's gone.** Don't record too many stereo tracks within a song. Final placement of stereo spread is determined during mixdown. When drums, bass, guitars, keyboards, horns, and vocals are recorded in stereo, the wide imaging of the individual instruments may get lost. I know, sometimes you wish the musician would get lost.

• **Don't buss single channels**. Route the signal directly to the multitrack recorder. A direct signal is higher quality because it has less processing.

• **Scan across the console to confirm that all busses are correct.** With all the signals routed every which way, one wrong button pressed could mean a great track ruined.

Documentation

• **What should the track sheet include?** The track sheet is, of course, the page that documents the date, all locations of contents, where the tracks are, names of specific files, and additional comments. All must be clear. It also includes tape speed, equalization curve, reference level, alignment tones, and recorder manufacturer. Also noted are take numbers, dates and possible small notes, such as microphones and outboard gear used. Cue lists and BPM are also commonly listed.

All paperwork must be clear enough so any other person can easily sort through it. Many times, a project uses any number of studios and recording engineers. Good, clean, complete and easy-to-understand track sheets are a staple of the recording environment and an important part of the session.

Information on parameters is also needed in the hard drive world of recording. Include notes about sample rate, bit depth, reference level, titles, and dates with the FireWire hard drive. This could be taped with a small card right onto the hard drive.

• **Mark the choice pass on the track sheet.** When naming a new track for the session, give it a proper label, not just 'guitar 7.' If a track is a comped master, label it with a *. If the track is in time, add a second *. If the track is in tune add *** to the end of the file name. For example, the final lead vocal was comped from lead vocal 1 through lead vocal 7.

With many freelance operators hired to add stars 2 and 3 to the sound file, a logical virtual notes system must be employed. When the project is finally ready to mix, all the choice tracks will have *** after their file name.

• **Labeling, labeling, labeling.** Some projects have hundreds if not thousands of files, including edited files, complete takes, partial takes, and unused takes. Place all these files in the master file list. Back up everything and remove any files that will not be used in the final mix.

• **Use a notebook for each project.** Keep the relevant information there. Loose pages get torn, mixed up and just plain lost. Personally, I am a fanatic at keeping paperwork. I have report cards from elementary school, love letters from Junior High, and that annoying restraining order thing.

• **How about a date?** Include the year when writing the date on anything. Years fly by, and, as anyone being audited knows, something with the date 'July 30' is pretty much useless.

• **Re-zero the counter for every song.** Keeping track of cues is easier when the zero starts at the top of the count-off.

• **My song is on the charts.** If your project has more than a few songs, make a large chart with all the titles on one axis, and all the instruments on the other. Finish a track then check it off on the chart. This makes it clear what has been done, and what needs to be done. Anyone can look across the room and see exactly where the project is at.

• **Write down favorite settings** that you stumble upon while tracking or overdubbing. Don't start from scratch on mix day. Experiment throughout the project. During the session, things are always happening. Often, something great might fly by, such as a great echo sound at one spot in a song. Writing a favorite setting down gives you more options by mix time, as you will have lots of ideas collected. If possible, record this great echo effect onto an open track. If it moves you, then it should also move the listener.

• **Department of redundancy department.** Conflicting, redundant, or just plain wrong information is worse than no information at all.

• **Keep to a schedule.** If your project is under great time restraints, make up a schedule that everyone can agree on. Post this on the wall, and stick to it.

• **Drum to your own beat.** Rather than blindly use the following settings, determine the characteristics of the musical instrument you are about to record. Every instrument sound has its own characteristic envelope.

Drums

Ask yourself a few questions before setting the processors. If bussing a lot of microphones to a few tracks, will you place a compressor/equalizer across the buss, or across each incoming signal? How many tracks do you have? More is not always better. Should the panning be set from the listener's perspective, or the drummer's? How much processing should be used? Should you record any effects with the signal? Combining tracks can be tricky until you can see where things will lie within the final mix, and set your levels accordingly.

• **Use lower levels.** Using a VU meter, set the recording machine input level of the drums low, occasionally in the yellow – rarely in the reds. Drums can have a lot of wallop and the transients don't show up on VU meters. Drums are often the pillars of a song, and must be sturdy enough to carry the load. Equalization and compression can bring out the natural crack and boom.

• **Peak monitor.** If available, use the peak meters to monitor levels of the drums. Don't overload the higher frequency instruments, such as high hats and cymbals. They have a lot of harmonics and high-level transients.

• **Use the proper drummer.** Have him play the song you are about to record. Getting sounds with a substitute drummer means nothing.

• **Pick a low snare.** When getting drum sounds, keep them at a reasonable level in the monitor mix. All drums sound great when turned up loud, but lower levels reflect a truer indication of how each drum sounds. If the drums sound full and clear at a lower level, imagine how great they'll sound when the volume is turned up.

Kick drum equalization

Proper Q settings can help define each drum by minimizing frequency overlap. Starting points might be:

– Pull below 40 Hz.

– Boost around 60–100 Hz to bring out the thud of the kick, maybe even lower in certain circumstances such as some dance mixes. Set the kick drum frequencies in tandem with the settings on the bass guitar. These two instruments carry the low end of the song and each should be distinct. Add a frequency on one and pull the same frequency on the other. Note that a tight kick skin won't have the low end of a looser skin.

– Pull around 164 Hz in the kick drum to bring clarity to the bass track. 164 Hz is a harmonic of the bass guitar's fundamental low E note, 41 Hz.

– Add up to 200 Hz for body and fullness. Watch overlap.

– Pull from 200 to 600 Hz or higher to remove unwanted cloudiness and to open room for other instruments.

– Boost around 2.5–5 kHz for solid thwack.

– Boost at 5–8 kHz for crispness, or a clicky sound. With faster tempo songs, the kick may need more click to be heard, while slower tempo songs leave room to allow solid lows to come through.

– Pull 8 kHz and up. These frequencies contribute little. Pulling them won't effect the sound much, and may reduce hiss.

Snare drum equalization

– Roll off up to 100 Hz to reduce muddiness.

– Boost somewhere between 100 Hz and 300 Hz for the body of the snare drum to come through.

- Boost somewhere between or around 500 Hz–1 kHz for that nice woody crack sound.

- Boost around 1 kHz for a 'tink' sound.

- Boost between 5 kHz and 10 kHz for crispness.

High hat equalization

- Roll-off everything below 180 Hz to remove rumble and leakage from other drums.

- Pull 500 Hz–1 kHz to remove 'clang.'

- Boost 3 kHz to add fullness or ring, but this is seldom needed.

- Find sheen at 8–12 kHz.

Tom-toms equalization

- Roll off unwanted low end. Some tom-toms can get pretty low.

- Find and pull the rumble on each tom-tom. Turn up the level control on the equalizer, but not a lot. Too much low end can carry enough power to blow out a speaker. Sweep the low frequency as the player hits the drum, and find any offending rumble, maybe somewhere within 300 Hz–1 kHz – the floor tom-tom might be lower. Find and pull the frequency using a narrow Q setting – not enough to radically alter the sound, since there is a lot of harmonic tone within that frequency area. As well, that frequency may have a higher harmonic sweet spot, so when you pull the rumble, it naturally enhances that spot.

- Adding lows, between 100 and 300 Hz, depending on the drum, will bring out the thud.

- Pull around 8–900 Hz to lessen the 'boxy' effect. Boost in this area to add boom.

- Use a narrow Q setting, and sweep between 3 kHz and 5 kHz to find the 'sweet spot' frequency and raise it. You'll know it when you hear it.

- Pull a bit around 8 kHz and higher to diminish cymbal leakage.

Cymbals/overhead equalization

- Roll off up 180 Hz to eliminate air conditioning rumble and low-level street noise.

- Boost a wee bit of 8–12 kHz to add shimmer.

- Sweep the high end to locate any frequencies that may need pulling or adding. The overheads preside over the whole drum kit, and tend to gel the individual drums together. A rich, full-sounding kit should have lots of rich highs, and few clashing lows.

Room microphone equalization

- Use the appropriate microphone roll-off to minimize rumble.

- Pull somewhere within 120–500 Hz to open up room for high priority instruments. The rich harmonics will still come through, with the power of the low frequencies coming from the individual drum tracks.

Drums compression

Only the very best drummers can hit the drum the same every time, so using compression is almost a necessity. Proper compression can bring up the lows and help deliver solid drum sounds. Commonly, drums can be compressed more than other instruments because they are less musical and more percussive.

Use these starting points:

Attack. 5–10 ms or maybe faster. A slower attack time can allow initial peaks to sneak through before the compression kicks in. This may give a nice added crack to a snare sound, but watch for overload. Set the attack time slower on the kick drum, as it may take a few milliseconds to build to its full potential. Fast attack and release times bring up the body of the drums and cymbals. An attack set too fast may diminish the initial crack of the drum.

Release. Start at 250 ms, then move to suit the song. A fast release time can bring up the level of the decay and raise the sound of the snares.

Ratio. 3:1 or 4:1. Drums, due their nature, have fast natural attack and release times, with plenty of peaks. A high ratio levels the dynamics while delivering the meat of the sound. Control the signal enough to record it, yet don't overcompress it so as to lose the initial transient crack. Of course, as the ratio gets higher, past 8:1 or 10:1, the compressor becomes a limiter. A limiter is great for eliminating transient overload on digital input circuits.

Threshold. Low. A lower threshold preserves the full impact of the drums, and can sustain the cymbal's natural decay.

• **Compressing the room.** Run the room microphones through a compressor on high compression and fast release. When the player is playing, the ambient microphone level is lower, removing any unneeded cloudiness. When the player stops, the ambient microphones open, making it sound as if the player is in a large ambient room. As with many things in the studio, compressing room tracks is your personal preference. Do what you feel works with the song.

• **Kick – the habit.** If your drum setup has two kick drum microphones, maybe compress the far one and pull a bit of the high end. Add it in with the close microphone to get some teeth to the sound. Or try placing a microphone on the opposite side of the kick drum where the beater makes contact with the skin. Switch the polarity on this microphone.

• **Level the drums.** If the drummer is playing with both sticks on the snare drum, some of the hits may be louder than the rest. To raise the level of the lower ones, split the signal into two. Affect one of them how you normally might, gating out all but the main or loudest hits.

Send the second signal to a limiter, and compress the loudest part. Send a buss output from the first snare channel into the sidechain of the limiter. When the snare on the first track hits, that controls the compression of the limiter. Blend the two tracks until both the crispness of the loudest snares and the subtle in-between parts are properly audible.

Drum gates

When placing microphones on the drums, complete separation is almost impossible. With proper microphone choice and placement, leakage from one drum microphone to an adjacent microphone can be minimized. Before you reach for the noise gate to eliminate leakage, choose the right microphone and place it properly. Messing with gates during recording can result in painting yourself into a corner. Sometimes it's better to wait and gate during the mix, especially when recording a dynamic drummer.

• **Listen to the gates.** If you are gating any of the drums while recording them, listen to the gated sound throughout the song to make sure the gates are being triggered correctly. Changing the input level of a drum track (sometimes even an equalization change) may alter the triggering of the gate. Check that all gates have triggered correctly before moving on.

• **Gate the drum.** It is not uncommon to add some degree of gating on the kick, snare and tom-toms, but the cymbal microphones are not normally gated during recording. If the internal trigger on a drum gate isn't fast enough:

(1) Mount a small contact (pickup or lavalier) microphone to the rim. This tight microphone will open the noise gate faster.

(2) Listen to the contact microphone, sweep the equalization to find the drum's most prominent frequency, and accentuate it.

(3) Run the signal through a tight noise gate to make it sound like a click.

(4) Plug this into the sidechain input of the noise gate on the drum.

This really only works when the player hits solid drum hits. A drummer using brushes for drumsticks, or lightly tapping the drums may not trigger the gates as planned.

• **Gate the tom-toms.** Eliminate leakage in the tom-tom tracks without using a separate microphone as a trigger. For each tom-tom:

(1) Split the signal coming from a tom-tom microphone into a second channel on the console. Insert a noise gate on the first tom-tom channel.

(2) On the second channel, determine the fundamental frequency of the tom-tom. Accentuate it by setting a thin Q, then pulling the other frequencies.

(3) Gate and equalize the signal so all that comes through is a solid click when the drummer hits the tom-tom. Leakage from any other instrument, even other tom-toms, should be dialed out.

(4) Run this output into the input of the sidechain of the noise gate that was inserted on the first tom-tom channel.

Any time the drummer hits the tom-tom, the trigger opens the noise gate, allowing the signal through. Due to the slow build of a tom-tom sound, try using a contact microphone.

• **Talk to the drummer.** Once you get the sounds to where you want them, record the drummer playing the song to be recorded, then have him come into the control room and listen. Once he is happy, you can move on.

• **Electronic drum machines commonly have 'hyped' sounds.** Electronic drums and percussion are already equalized and compressed, so some of the previous equalization and compression settings may be unnecessary.

Bass Guitar

Bass guitar equalization

Because recording bass guitar commonly falls in the realm of basics, it can be difficult to know what frequencies to add or pull until all other tracks are recorded. Many engineers record the bass with no equalization. During the overdubs stage, they play around with different settings to find what works best in preparation for the final mix.

• **Get solid.** The bass must be solid. Its job is to hold the low end together, so any level fluctuations by the player must be made uniform by using filters, equalization and compression. As a starting point:

–   Dial out the very lows, maybe up to 40 Hz, using a high-pass filter so the compressor doesn't have to deal with unneeded rumble.

–   Add between 50 Hz and 80 Hz for real lows on the bass. But don't mush this area with the kick drum. Add (or pull) a frequency on either the bass or the kick drum, but not both.

–   Add around 82 Hz to add weight and to keep the bottom of the song solid. The punch dwells in these low frequencies.

–   Use a narrow Q setting, and pull between 160 Hz and 350 Hz for clarity.

–   Add somewhere between 600 Hz and 1.5 kHz for the nuts to come through. Pull here and raise the overall level to open space for other instruments. Adding too much here can make the bass guitar sound vague.

–   Add 2–2.5 kHz for presence, but watch overlap with other instruments.

–   Add around 3 kHz to bring out click and fret noise.

–   Pull from 4 kHz on up for a smoother, less attacky sound.

Bass guitar compression

Often the bass guitar has fewer initial transients than an electric guitar, again depending on whether the player uses a pick or not, whether he uses a 'thumb slap' style, or whether he uses his fingers. Different amplifiers require different approaches. A valve (tube) amplifier has a natural musical compression, so additional compression may not be necessary. A close miked setup will need different settings than a distant setup. Dynamics play a major role in setting levels, because a more dynamic part will need different settings than a smooth part. As a starting point, try:

Attack. 20–60 ms. A slower attack time may help define the bass sound so that the initial note passes before the compression kicks in. Due to the low frequencies involved, it takes a bit longer to reach maximum power. Go faster, 0–20 ms for a 'clickier' sound. For a smoother transition from non-compressed to compressed, try pressing the soft knee button.

Release. Medium to long. Use a longer release for substantial sustain.

Ratio. 4 to 1. It depends on how wild the player is – a good bass track shouldn't need a high compression ratio. Adding too much compression may affect the higher harmonics of the bass guitar.

Threshold. Low. Maybe -25 to -15. Again, it depends on how smooth the player is. If he keeps his levels uniform, a higher ratio may work. A low threshold with a high ratio can result in a nice sustain. A ratio that's too high, coupled with a threshold that's too low is noisy and can suck the dynamics from a track.

• **All strings considered.** Sometimes you must control certain frequencies, and not others. Maybe the bass guitar has one note louder than the rest, but pulling the specific equalization could degrade the sound. Try this:

(1) Return the bass to another channel.

(2) Sweep the frequencies to find the offending note.

(3) Pull the rest of the frequencies, except for this note.

(4) Patch this equalized signal into the compressor's sidechain input. The sidechain dictates the compression, so only the input note is affected. This channel is for processing, and not part of the audio signal.

• **Sneaky peaks.** An overly dynamic bass track may need a compressor plus a limiter to catch any peaks that sneak through.

• **Know where you are going.** Determine what you want before you start twiddling. Don't change things hoping to find some magical sound. Picture it in your mind and then dial it in.

• **Does the player play with a pick during the song?** Check beforehand if the player uses fingers or a pick. Sometimes he may use, for example, a fingering method during the verse, and a pick during the chorus. This means the settings on the processors might need changing to compensate.

Electric Guitar

Electric guitar equalization

Of course, you don't really equalize the guitar, you equalize the microphones on the guitar amplifiers. With such a wide array of guitars, players, and styles, it is impossible to say there is a right or wrong guitar sound. Once the setup is complete and the microphones are correctly placed, have the guitar player play the part to be recorded and, as a starting point, try:

— Rolling off below 80 Hz. There is little musical information from a guitar lower than 82 Hz so adding will just add unneeded low resonance. Leave this area for the designated low-end instruments: the kick drum and the bass guitar.

— Adding between 80 and 250 Hz for extra fatness. This area contains the fundamentals of the instrument. Maybe pull here if the guitars are not the primary musical instrument, or if they overlap with the vocal.

– Adding between 250 Hz and 1 kHz for content. This is an important guitar area because the second and third harmonics of the A and D strings live here. Adding the right frequencies in this area can place the guitar more in the foreground. Pulling here places it farther away and makes room for vocals and other instruments.

– Sweeping the upper midrange, between 2 and 6 kHz for the spot that has enough bite. Somewhere there is a spot where the sound isn't too brittle or too boomy, and has just enough distinction.

– Adding 5–8 kHz for added richness, but this can also make any buzzes more apparent. Due to an amplifier's limited frequency range, musicality decreases with higher frequencies.

– Adding some very highs, over 8 kHz can give an electric guitar some air, but can easily turn the sound brittle.

Electric guitar compression

On a loud electric guitar amplifier, there is lots of low frequency harmonics and overtones, but not a lot of highs. Take proximity effect into consideration while setting compression levels. Too much low end can dominate, then dictate the compressor's reactions.

Medium-loud guitars may need less compression as the proximity effect on the microphones may not be as apparent. Try:

Attack. Semi-fast to fast, such as 10–50 ms. Semi-fast attack times can smooth out a lead guitar solo.

Release. Long. From 0.5 to 1 second. Longer release times heighten the sustain. Shorter times increase pumping, especially with high ratio and low threshold. Start by setting the release where it returns to zero just as the next beat of the song hits.

Threshold. Low. Start at -25 and slowly raise it so all the crunch is brought to the surface. Unfortunately, this also brings up the noise.

Ratio. At least 5:1, maybe as high as 10:1 for sustained guitars. Use the peak level control, if available, to keep the initial attacks at bay, especially if the attack setting is at all slow.

• **Compensate at the amplifier before the control room.** Once you get the sound you want, compare the settings on the amplifier with the equalization at the console. If the amplifier has, for example, certain frequencies pulled, and you are adding them at the console, set the amplifier settings to compensate then remove the console equalization. Get the best sound from microphone placement and the tone controls on the amplifier. Always ask the player before touching or changing his amplifier setting.

• **Use distortion as a creative tool.** Use distortion to add edge to a sound. Tube distortion works best when the tubes are being driven, so turn up that tube amplifier. But everything distorts differently. Digital overload distortion will always sound terrible.

Acoustic Guitar

Acoustic guitar equalization

Properly miked, a quality well-tuned acoustic guitar with new strings should need little equalization – perhaps to add some frequencies for sheen, or maybe to pull where the sound may mask other instruments.

When equalizing something with as many overtones as an acoustic guitar, pull the unharmonic overtones and enhance the pleasing harmonics. Here, the musical recording engineer has the advantage. As a starting point, maybe:

– Roll off below around 82 Hz. The lowest note on the standard acoustic guitar is E, around 82 Hz.

– Sweep the low-midrange, from 80 Hz to 300 Hz to find the boomy sound, then pull it using a narrow Q setting.

– If there is room, add somewhere between 80 Hz and 350 Hz for body.

– Add 300 Hz–1 kHz for early harmonics.

– Add somewhere between 700 Hz and 1.2 kHz for more 'wood' or pull here to ease the secondary harmonics.

– Add 1.5–3 kHz for presence. Pull for hollowness.

– Add 3–5 kHz for presence and attack.

– Add 5 kHz–10 kHz for brightness.

– Add around 10–12 kHz for sparkle. However, it doesn't take much to go from sparkle to brittle. Adding highs means adding noise.

Acoustic guitar compression

The characteristics of an acoustic guitar might include wide dynamic range, semi-fast rich initial transients and substantial sustain. The acoustic guitar may not have as many peaks as a snare drum, unless the part is percussive, but it does have peaks. Closer miked sounds may need more compression than microphones placed a couple of feet away. Try:

Attack. 10–20 ms. A very fast attack can control the initial transients of a sound.

Release. Medium. Start at 250 ms and raise or lower as the tempo of the song dictates.

Threshold. Medium to high. A high threshold allows all the natural sounds and dynamics of the guitar to remain intact. A lower threshold might bring out more lower body.

Ratio. Low, to begin with, maybe 2:1 or 3:1. A higher compression ratio may be needed as a player may tend to move off-axis now and again. Play with the ratio until the quiet bits as well as the loud bits can be heard. A higher ratio can increase the sound's density so it fits in with the other compressed tracks.

• **Choose to use two.** If you choose to use two microphones on an acoustic instrument, the one with more lows – usually the closest microphone – may need more compression than the distant one.

• **De-ess the guitar.** Minimize fret squeaks and noise with a de-esser.

• **Defeat the proximity effect.** Pull some of the low frequencies that may be created by proximity effect so the compressor won't react to the added lows.

• **Sympathy for the level.** When the acoustic instrument is not in use, put it away, or loud levels in the room will cause it to ring out sympathetically.

Vocals

Vocal equalization

• **Eek. You EQ?** Better engineers prefer microphone choice and placement before equalization to get their vocal sound. If you MUST equalize, have the singer sing the song and:

– Pull up to 80 Hz. This area can be rolled off for most vocals as it holds all microphone rumble and not a lot of clarity. Pulling these lows on the vocal will open space for other instruments that benefit from this frequency, such as the kick or the bass. Try a microphone roll-off instead of an equalizer.

– Add 80–300 Hz for thickness and body. This area contains the vocal fundamental tones of many singers, sometimes higher for a female voice, so it carries a lot of the weight of the vocal sound. Tread lightly in this area. Pulling these frequencies can result in a hollow boxy sound. Adding here risks masking other important instruments.

– Pull around 300–500 Hz depending on the voice, to ease offending first harmonics. Adding too much in the 500 Hz–1 kHz range results in a honky nasal-like sound.

– Add 800 Hz–2 kHz for intelligibility. This area adds warmth or quality of voice. Pulling some of these frequencies can eliminate harshness, but too much can result in a hard, colder sound.

– Pull around 1 kHz to allow space for other instruments. This works well on background vocals as they are commonly used as support rather than the main focus.

– Add 2 kHz–6 kHz for clarity and presence. This where the s's and the t's cut through. Too much in this range and a de-esser may be needed. Pulling here makes the vocal sound dull. Adding too much makes the vocal sound shrill.

– Add from 6 kHz to 16 or 18 kHz for sibilance and air. Setting the right equalization in this area can place a vocal just perfect, but mind the hiss created by raising these high frequencies.

Vocal compression

The characteristics of vocals, depending on the singer and situation, can include fast attack and wide dynamic level changes.

Some vocals need limiting and compression, but some vocals don't. Some engineers love to compress all vocal tracks, while others choose to record vocals with as little processing as possible. Maybe choose selective processing, using compression only on the needed bits. Do whatever works, as long as the louder parts don't overload and the quiet parts stay out front. The key to good vocal compression is subtlety. You don't want to hear the compressor working, you just want to hear every note and nuance.

Compression settings on the vocal depend a lot on the settings on the rest of the tracks. A rock track may have tighter overall compression settings than a love ballad. The vocal must stay buoyant among the rest of the tracks, so settings must be uniform with the rest of the instruments. Try:

Attack. Fast to medium fast, depending on the tempo of the song. Too fast an attack time can mask the very beginning of the vocal line. A soft knee setting on the compressor works well on vocals as it smoothes the transition into compression.

Release. Set the release time at fast for a faster song, and set it at slow for a slower song.

Threshold. Start at the highest, then work backward. When you see the needle move – stop. You want the loud passages to compress and the regular full voice to move the needle minimally.

Ratio. Start with 3:1 or 4:1 depending on the singer and the part. Get a solid output, then set the ratio for the best final settings.

• **Set the settings according to his abilities.** Good singers have control over their dynamics and know how to work a microphone. They might not need as much compression or limiting as a less experienced singer. Bryan Adams is the king of singing loud with no microphone overload just by knowing when to move off-axis a bit.

• **Set the settings according to the song.** Determine the dynamics of the vocal part. If the vocal is consistent in level throughout a song, the settings would be different than if the vocal were more dynamic in certain sections. Very dynamic tracks might benefit from limiters over compressors, as a properly set limiter will not allow the signal to overload.

• **Place a peak limiter after the compressor.** A dynamic, close miked vocal can have such seriously high peaks that one compressor alone may not be able to control them.

• **The background vocals have more compression than the lead vocal.** Background vocals are often not as dynamic as the main vocal, so they can be more compressed to sit nicely in the background. For lead vocal clarity, buss the lead vocal into the sidechain of the background vocal compressor. This automatically lowers the background vocals whenever the lead vocal is singing.

• **Limit when you record, and compress when you mix.** Limiting a pre-recorded vocal is not like limiting a live vocal. Due to the limitations of analog recording equipment, a pre-recorded vocal can't be as dynamic as a live vocal. A digital recording can preserve the original dynamics.

• **De-ess the vocal.** A de-esser placed first in the chain controls any sibilance problems before reaching the compression and equalization. The de-esser can eat some of the needed sibilance if it's set too high. Try waiting until the mix to ease out the s's, as long as the vocals are recorded properly.

• **I auto-pitch it.** Many engineers connect an auto-pitcher in the signal path in case the singer's pitch strays a bit. Record the vocal normally, and record the 'tuned' vocal on another track. Then, if the tuning on the first track goes way out, bounce only the needed bits from the tuned vocal track.

• **Gating the vocal is uncommon.** Gating the vocal microphone may well be common in live settings, but studio situations are more intimate. All the vocal inflections must be heard. A quiet note is no less important than a louder note.

Horns

Horn equalization

Horns can be particularly brittle if recorded incorrectly. Again, microphone choice and placement have a huge role in getting the best sound. Poorly written horn parts will be harder to record. Maybe:

– Unless the song relies on horns for the bass parts, install a microphone roll-off to pull all the frequencies below the horn's lowest note.

– Add 200–500 Hz for the body to come through, but again, depending on the placement in the final mix, you may not want the horns masking the vocals.

– Presence at 2–4 kHz. Watch overlap in this area. This is a good area to pull when the horns aren't to be feature instruments.

– Boosting from 5 kHz on up can bring out brilliance, but too much and the sound turns shrill and brittle.

– Boosting 8 kHz and up brings out click from the keys.

Horn compression

Horns can be dynamic and loud, with high initial transients and little decay. Too much compression can remove some dynamics and thin out the sound. Compression may not be needed on all the horns, maybe just the lower horns such as baritone sax. Higher horns tend to cut through naturally, so they may not need the compression.

Attack. Fast. For many horn sections, usually as fast as possible.

Release. It depends on the song. If the horn parts are choppy, keep the release faster. A solo horn might have a medium release time. Horns being used as pads might have a slower release time.

Threshold. Low, perhaps -10 to -15 dB. A lower threshold can pull up all the nice growl.

Ratio. 3:1. This is a standard starting point. Raise it a bit to control the low end, especially if the player likes to move around a lot. Keep the ratio low, as horn players tend to 'swell' a lot as they play. Higher ratios level these emotional swells. Just don't overdo it.

• **Flugel. Double Flugel.** Doubled horn tracks sound great, except for low horns. The low-frequency area needs clarity, and that area belongs to the bass guitar and kick drum.

Piano

Piano equalization

A piano has a rich full sound, initial high transients and lots of natural decay. The sound depends on so many variables, including placement, microphone choice and placement, player ability, song writing, and so much more, that these starting points are of little use.

147

If you must equalize the piano, use a wide Q setting with minimal level changes. Adding too much equalization on the piano results in a mix where the only piano frequencies you hear are the ones you added. As a starting point:

– Roll off the very lows, below 50 Hz.

– Bring up the low end by boosting 80–150 Hz.

– If the sound is too boomy, pull frequencies in the 200–400 Hz range.

– Add presence and attack by boosting between 4 and 8 kHz.

– Add air by adding 8 kHz and higher.

Piano compression

Start by setting the compressor as a limiter with a high threshold and work backwards from there. As the player plays the song to be recorded, set the levels to where the signal moves during the louder bits.

Attack. An up-tempo song may have a faster attack than a slower piece. Set the attack according to the tempo of the song.

Release. A medium release time is a good starting point. As with the attack settings, fine tune it once the tempo of the song is determined.

Threshold. A lower threshold means more of the sound is compressed, and the piano has a full sound, with a lot of fundamentals and overtones. Set too low, overtones lose distinction. A higher threshold might be a good starting point.

Ratio. A high ratio ensures the signal will not overload, even if the player decides to go all 'Jerry Lee' at the end of a smooth love song.

Something's not Right

Sometimes you simply will not be happy with a sound. You've changed microphones, you've equalized, you've compressed, you've put it in a taxi and sent it round the block – you've tried everything. Maybe:

– Remove all processing and listen to the track within the core instruments, often bass and drums.

– Turn any amplifiers down or up. Listen at different levels.

– Disconnect any or all pedals in the signal line.

– Use a designated heavy-duty cable from the instrument to the amplifier, or direct box.

– Have the player play the song you are working on.

– Tweak the controls until the sound doesn't stink. Don't spend an hour on it. Pull a bit of the midrange frequencies to clean up any muddiness.

– Change the equalization on the other instruments so it fits in better. Sometimes equalizing the other instruments that surround a track will give it more definition.

– Have your assistant move the microphone a bit to get a better sound off an amplifier or instrument.

– Try different volume levels at the amplifier.

– Change the amplifier or, even more drastic, the tubes inside the amplifier.

– Turn off all pedals or effects, then bring them in one at a time.

– Listen to the sound in the track, with the vocals.

– Wait. The tone will improve as the player gets warmed up.

Chapter Nine

THE SIGNAL ROUTING

In the analog recording world, recording too hot results in distorted sounds. In the digital world, hotter means cleaner sounds because you are using more of the 'volume bits.' But when signal reaches 0 dBFS, sounds will instantly become 100% distorted.

Lowering the recording levels would avoid overload, but lower it too much and signal leaves the high 'hot' area, or digital MSB – Most Significant Bits – and edges into the lower resolution of LSB – Least Significant Bits.

• **Set the microphone input fader at zero.** This is not off, but about 3⁄4 of the way up. Set the record input level with the microphone input trim control to take advantage of the console's optimum parameters.

• **What is trim?** The trim knob is an active gain control within the pre-amp. Because different instruments and microphones can have vastly different levels, the trim lets you raise and lower incoming signal for optimum level.

• **Avoid overloading the system.** Lower your input levels for high transient tracks such as percussion instruments.

• **Lovely read a meter aid.** When recording musical instruments with high transients, rely on your ears and your experience, using the meters only as an aid. Remember, peak levels tend to be too fast for the VU meters, and might show inaccurate readings.

• **Players will play harder as a song progresses.** Perfect levels at the start of the song may creep into the overload zone by the end of the song. Set input levels a touch low before pressing the record button.

• **What is headroom?** Headroom is the distance, measured in decibels, between a device's optimum operating level and the distortion point. More headroom can mean more room for peaks.

• **What is signal-to-noise ratio?** This is the ratio, measured in decibels, between the noise floor and the optimum signal level. A higher number is desired here so more information can be recorded.

Levels to Record

• **Keep it quiet.** Just like links in a chain, the console is a linked series of circuits. If one link is low, boosting somewhere else to compensate adds unnecessary noise and distortion. A little bit of noise here and there adds up to a whole lot of noise in the final result. On rock records, there may be a bit of leeway, but on jazz, classical, or solo instrument recordings, unnecessary noise is unacceptable.

• **Get the best level as early as possible in the gain stage.** The sound recorded will be cleaner, with less hiss and noise. A louder signal means less additional level is needed at the pre-amp stage. A low level will mean raising gain control resulting in increased noise. Levels that are too hot will overload the inputs.

Take the needed time to set the optimum operating level for the cleanest sound and the least noise and distortion.

• **Good vs. bad.** Every device introduced to the signal degrades the signal's definition. Less processing means purer sound, with more focus and clarity. The best signal has the least amount of processing. You can dumb down a good signal, but you can't smarten up a bad signal.

Figure 9.1 shows good gain staging. Minimize noise while maximizing gain at:

(a) The input stage. The right microphone enters the console at the proper level. From there, the pre-amp raises the input level. No pad or trim change is needed.

(b) The output stage. Compression and equalization are not connected into the signal path, and a healthy level is going 'direct' to the input of the recording machine.

(c) The monitor stage. The clean signal level then returns to the master fader at 0 VU.

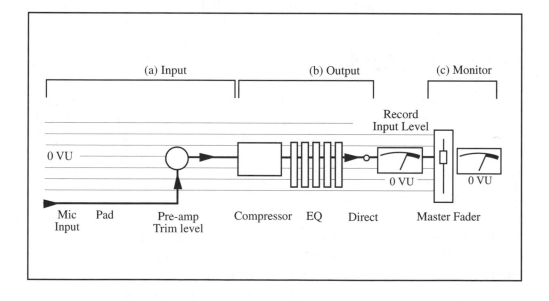

Figure 9.1 Good chain

Figure 9.2 shows an unnecessary amount of noise introduced into the signal chain at:

(a) The input stage. The microphone level is too hot for the input, so the pad is introduced. This lowers the signal level into the pre-amp so the trim level must be raised.

(b) The output stage. The input level to the compressor is too hot, so the input is lowered. To compensate, the output level is then raised so the signal level overloads the next component, the equalizer. This overloaded signal level is then lowered at the buss, and this low level signal is sent to a recording machine input.

(c) The monitor stage. A signal recorded with low level means the fader must be raised higher than normal for the signal to reach 0 VU.

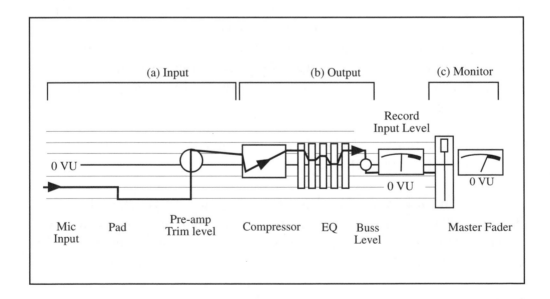

Figure 9.2 Not so good

Monitor

• **Monitor the return from the recorder, not the channel input.** Before setting levels and sounds, route the signal from the microphone through the console to the recording machine, and back to the monitor section at the console. By not monitoring the output of the recording machine, you don't really know what's there. An errant buzz or click might slip by unnoticed.

Monitoring the returns from the multitrack recorder allows you to stop and punch in whenever the situation calls for it. If you are monitoring the inputs when you record, you won't hear the recorded track to punch into.

• **Get fast.** Good engineers can get great monitor mixes in just a few moments because they know what they want. This isn't the time to do any kooky equalization or wild effects. Just quickly set up a solid monitor mix so the recording can proceed.

• **Pan the players.** In the monitor mix, place the players as they are in the studio, including the drums as you see them. For example, if a guitar player is toward the left side in the studio, pan the guitar to the left side in the control room. If the high hat is on the right, then pan the high hat to the right.

If you choose to pan the drums from the drummer's perspective, where the high hat is on the opposite side, that's fine. Just pan the cymbal microphones the same. Don't, for example, pan the high hat from the drummer's point of view, then the cymbal microphones from the listener's point of view.

• **Use the mute buttons.** To eliminate unwanted console noise, make a habit of muting all the unused channels.

• **Flange can mean you're hearing it in two places.** If you hear the sounds 'flanging' while monitoring the output of the recording machine, perhaps you are hearing the input as well as the output. Maybe the channel sending signal to the multitrack recorder is accidentally being sent to the monitor buss too.

• **Don't send everything to everything.** Pick and choose which effects works best with the music. Continually work out your layout and direction on the rough mixes, listening to which parts pan best with which parts, or what grooves work with what level changes.

• **Remember Fletcher/Munson.** Another reminder to monitor at low volume levels, not only for your ears but for your sounds. Fletcher/Munson curves chart equal loudness points, and state that different frequencies will appear louder or quieter at different levels. This tells us that what may sound great at volume 9 may sound bass shy at volume 2.

Cue Mix

• **Ignore everything else but this.** For the best performance, set a great headphone, or cue mix. If a musician is really stoked with the sounds, the cue mix and the studio vibe, the performance will improve. Take the time to get the absolute best mixes for everyone.

Go into the studio yourself and listen to each cue mix. Speak into the microphone, maybe sing along. Have the band play, and listen. Can you hear everything? Is the vocal going to be loud enough? Too loud? Are the effects as you want them to be? Are left and right sides equal? Does it sound rich and full in the headphones?

• **I see cue.** Determine whether the cue mix should be set at pre-fader or post-fader. A pre-fader cue mix lets you to change the fader levels in the control room without changing the headphone levels. A post-fader cue mix means headphone levels change as the fader levels change.

• **Get the proper level for the vocals in the cue mix.** If the singer's vocal track is not loud enough, he may either slowly move closer to the microphone as the session progresses or worse, to make up for low levels, sing beyond his comfort range. This may result in a musically sharp vocal track. Just the opposite may happen if the vocal is too loud in the cue mix. He will not sing to his comfortable level, and may sing slightly flat.

• **Buzz off.** Tell the player to leave the headphones off until you are ready to press the record button. No one needs to hear the clicks, pops and feedback that can accidentally shoot through headphones as signals get routed. Only when you have checked the headphones yourself, and are finally ready to press the record button should the player don the headphones and begin.

• **Don't turn the cue all the way up.** Set the overall cue mix level from 1⁄2 to 3⁄4 of the way up, so you still have plenty of room for more volume. Loud enough so they can hear it, yet not so loud it will blow their hearing out. Better to hear 'Hey, Buddy, turn the cue up' than 'Hey, Moron, the headphones are too loud.'

• **Comfortable is best.** Just as the players were set up in the studio as they normally play with each other, so should the cue mix. For example, if the drummer is used to hearing the rhythm guitar to his left, set up the cue mix so that's how he hears it. It may be vexing to see a player on his left, while hearing him on the right side of his cue mix.

• **The cue mix is a priority,** not something to be set and forgotten. Continually check that each cue mix is the best it can be for each player.

• **Avoid abrupt changes in the cue mix,** such as level changes or effects return levels. If you change something on the console and it accidentally goes to the headphones, it throws the player's concentration.

• **Bad news, you both have mono.** If you only have one stereo cue send, but need two mixes, split the stereo send into two mono signals using the pan and volume at the console to get the best mono mixes. Of course, stereo mixes always sound best.

• **Use a dedicated reverb for the vocal.** At any time of day or night, whenever he wants to do a vocal, the effects in his headphones are set exactly how he likes it.

• **Set up out-of-phase speakers.** For vocal overdubs, if headphones are too uncomfortable for a singer to wear, Figure 9.3 shows the opposite polarity speaker setup.

(1) Place two speakers in front of the singer at ear level.

(2) Make a triangle by placing the microphone directly between the two. Use a tape measure to guarantee both speakers are equidistant from the microphone capsule.

(3) At the back of one of the speakers, switch the two wires so the polarity is opposite to the other.

(4) Send a mono signal to the speakers. Because their polarity is opposite, compression from one speaker (a) and rarefaction from the other speaker (b) will combine and cancel at the microphone.

(5) Play the song and listen to the microphone. Often the leakage heard is no worse than leakage from headphones.

(6) Although minimal, there is leakage. Don't play a track in the cue mix that you don't plan on using in the final mix. Leakage from a track that isn't there can come through loud and clear when you turn the vocal up during the final mix.

(7) Of course, this works only during overdubs, not when the whole band is playing along.

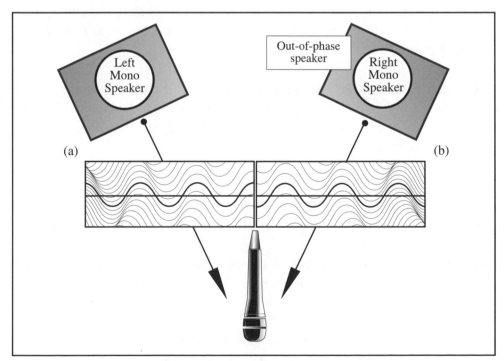

Figure 9.3 *Opposite polarity speaker setup*

• **Clearly label the instruments on the cue stations.** Larger studios today use individual eight channel consoles at each cue station. Perhaps the drums return on two tracks, then bass, acoustic guitar, etc. return on the remaining channels. When the player wants a louder track in his headphones, he just has to turn it up. Everything should be crystal clear for the musicians, so if they need more acoustic guitar, they can just reach over and turn it up.

But few players are recording engineers, so a player may hear something is wrong, but not know exactly what. Go out into the room and set up everyone's cue station to their needs. For example, the drummer may need to hear more click track, or the bass player may need to hear more kick drum.

• **Don't force anyone to wear headphones.** Players not too experienced in the studio may not want to wear cumbersome headphones. Bands rehearse without headphones all the time, so all of a sudden forcing a player to wear headphones might make him uncomfortable. To get the best performance, the players must be at ease.

• **Keep the drums low in the headphones during basics.** Everyone in the room can feel the drums because they are usually playing right there next to them in the room.

• **Give the drummer his own cue mix.** The drummer may want louder levels, and maybe no one else wants a loud click pounding away in their headphones. Other players in the room may not want any click in their headphones, but want to play along with the drums.

• **Keep it simple.** During an overdub, mute some of the secondary instruments in the cue mix. With fewer guidelines a player may come up with a few ideas that wouldn't come out if all the other instruments are there.

• **Keep the effects on the vocal in the cue mix to a minimum.** The singer needs to hear his own voice. Chorus messes with his tuning, and delay messes with his timing.

• **No fooling.** If the player wants a change in the headphones, do not try to fool the player by pretending to turn up a track, or pretend to change a setting. This embarrasses him, and makes you look like an amateur.

Chapter Ten

THE RECORDING

You've read it a million times, but it's true. There is no right or wrong way to record. That doesn't mean it's okay to make a mistake. That means you can be as creative as you want, just record the tracks right. Before recording, run through the process one last time in your head:

– All recording machines are cleaned, aligned, professionally maintained and ready for use.

– All the appropriate microphones are placed in the appropriate locations.

– The instruments are properly set up.

– The players are somewhat conscious.

– All microphones and headphones have been checked.

- All signal flow to and from the recording machines is correct, and the applicable tracks are in record-ready mode.

- Digital tapes and drives are formatted, exercised and properly cued.

- The console has been checked for routing, processing, monitoring, and cue mix settings.

- There are no external noises such as noisy refrigerators, heaters, fans or air conditioners, and all doors are closed tight.

- Any click track or references for the players are low in volume and ready.

Then you are ready to press the record button.

Tuning

• **Everyone uses the same tuner.** When all the players use the same tuning reference, all instruments will be in tune with each other. Imagine how it would sound if each player tuned to a different reference, then all tried to play together. Probably a lot like my band.

• **Buss to the tuner.** Use a buss to send the signal from the instrument into the tuner. If, for example, buss 16 on the console is open, patch the output of buss 16 into the tuner, then press buss 16 on any channel to route it into the tuner. This allows the players to check their tuning with no change in setup.

Some tuners have a 'through' output plug point that places the tuner into the signal chain. Don't use this connection in the studio because the tuner's internal electronics may degrade the sound of the instrument. They are great for rehearsal, but not for recording.

• **Record a tuning tone.** Better tuners allow the user to set the calibration with a pure tone, such as a tuning fork, or a single note from a guitar or keyboard. Once the tuner is set, record the note. Then if the tempo of the song gets changed, the tuner can be properly calibrated at the new pitch.

• **Mute all effects during tuning.** The main features of chorus programs are to introduce slightly different pitches for a wider, thicker effect. The player needs to hear only the instrument, not a pitch changing effect.

• **Having a hard time tuning to the track?** If you must tune an instrument to a pre-recorded track and there is no tuning tone, run the pre-recorded keyboard or acoustic guitar track into your tuner to align the tuner. Maybe the sustain of a guitar at the end of a song is enough for the tuner to read. Then tune the instrument to the now aligned tuner.

• **Tune an instrument only with the musician's permission.** Better to leave the instrument alone and let the player tune it. If you tune a guitar and a string breaks, it's your fault. Let the player break the string.

• **Don't play anyone's instrument.** Are you auditioning for the band, or are you the recording engineer?

• **Assault with a medley weapon.** Remind players to check their tuning on a regular basis. Out of tune recordings make people anxious, and makes the final product sound amateur.

• **Louder volume levels mask slight pitch problems.** Turn the studio monitors down to help hear pitch and tuning issues.

• **Eliminate all hums and buzzes before pressing the record button.** No gate, equalizer, or quantum gizmoizer can remove unwanted hum. You may think you are saving time, but time saved is forgotten when this buzz pops up every time you play the track.

• **What is the VSO?** The Variable Speed Oscillator changes the speed on the tape recorder. The VSO changes both the tempo of the music and its pitch. Hard drives don't have a VSO, as they alter the tempo and the pitch digitally.

• **Turn off the VSO.** Remove the varispeed from all recording machines before pressing the record button. If you unknowingly record a song with the VSO set at a different speed, then the next time you play the song back on the machine with the VSO off, it will be the wrong speed – and pitch.

• **Use the VSO for more bottom.** To get a large drum sound with lots of lows and thud, record a click track at the song's normal speed. Turn up the VSO and record the drums playing the song at the faster tempo. Remove the VSO, listen back to the recording and voilà – thick full kicking drums at the correct tempo. Normal sounding drums will have more lows, bottom and general boom, some of which may need to be equalized.

A higher speed on the VSO means more high-end frequency loss when played back at the correct speed. The opposite effect happens when the VSO is set slower. Drums played back at normal speed sound bass shy.

• **Note to self.** Place a large note on the VSO button to remind you that it is activated. This keeps you from accidentally recording the next track at the wrong speed.

• **Get it right.** Get the sound right before you press the record button. A 'fix it in the mix' attitude is a distraction that takes too much mix time and energy.

In Record/Red Lights

• **Record everything.** You are the recording engineer, so record. You have the option to redo something, but if some musical spark hits and then is gone, everyone looks at you to see if you recorded it.

If the player is messing around with a sound, or trying out new ideas, or even waiting for you to finish something, record him on a blank track. Less experienced players sometimes come up with ideas more readily if they think they aren't being recorded.

• **Keep track of all signal paths.** Continually check that all the equipment is acting as intended.

• **Attention all toe-tappers.** Don't worry if you can hear a singer or player tapping their toe to the music as they record. The taps are usually along with the beat of the drums, and will be masked when the rest of the instruments come into the mix.

• **Change an input level on a downbeat.** If the level of a track being recorded must be adjusted, do the adjustment on the downbeat of a change, such as when the verse goes into a chorus. If this shift in level is audible during the mix, you will know exactly when to make the change because it is not at some random spot that is difficult to pinpoint.

• **Ride the player.** If a player changes drastically from soft to really loud throughout the song, either:

–   Compress it more than usual.

–   Ride the track. This means to watch the player, and when he goes into overload mode, physically lower the fader to manually level the louder parts. This is a tricky bit of engineering, and can be fatal if you ride the level too much, making repairs in the final mix that much more difficult.

–   Re-record the softer parts after the louder parts are complete.

–   Use two separate microphones, one for the loud parts, and one for the softer parts. Record them both on separate tracks.

• **Don't stop the players unless you must.** Musicians don't like to be stopped when they are playing, so wait for the end of the piece.

• **Avoid click track leakage.** Lower the level of the click at the end of the song to keep it from leaking into other tracks as the instruments ring out.

If the player wants the click track really loud in his headphones, lower the high frequencies on the send before raising the level. This may reduce some unwanted leakage.

• **Take five.** If you are doing lots of takes, slate all the takes. Keeping track of counter numbers is great, but actually recording the words 'take thirty seven' before the count-off leaves absolutely no doubt as to your location. With digital recording, you have a visual readout on the screen, so slating might not be needed. Patience would.

• **Changing from one song to another should be fast.** If the tracks are recorded similarly, all the effects used on the previous song should be set and ready to go. With a little minor tweaking, such as a tempo change on the delays, you should be ready to press the record button.

• **Short of a technical problem, keep anything from bogging down a session.** Don't spend an hour getting just the right reverb sound for a minor overdub. Keeping the vibe up means keeping the creativity flowing throughout the session.

• **Take a picture, it lasts longer.** Computer snapshots are great to log those times when everything just seems to sound right. Often during overdubs and editing, while trying out mix ideas, a verse or chorus sounds perfect. A snapshot will preserve this idea until mix time.

• **Save a strip for me.** Save a favorite equalization setup by matching it to an out-of-the-way channel. For example, you really like the equalization settings on the guitar microphone, and you want to use the same settings on another song later – but the guitar channel is returning to the middle of the console. Rather than leave it there and take a chance that you may accidentally change it, move the settings and any patches used to an unused channel at the far end of the console.

• **Space out.** Leave enough space between songs on the multitrack recorder to be able to record a long intro of a song. This leaves the option open to really let an overdub ring out at the end of the song without the fear of going into the intro of the next song. Plus more time at the intro will allow for any future synchronization.

When recording to hard drive, organize passes by setting markers at, for example, 10 minutes, 20 minutes, 30 minutes, and so on. Start first pass on marker 1 at 10 minutes. On the first pass, maybe the first chorus is at 11:04. This will allow you to quickly go to any section on any pass. The first chorus on take three is at 31:04. The first chorus on take nine is at 91:04.

• **Run a second recorder,** a cassette recorder, even a portable boom box between takes to catch all the small talk, ideas, jokes and chatter that most bands do. Maybe the Philharmonic doesn't, but new bands are always excited to be in the studio, and often come up with some great fresh ideas on the spot.

• **Anticipate what is coming next, and quietly prepare for it.** Work like a duck on a pond. He looks smooth on the surface, but is paddling like crazy just underneath to get the job done.

• **Relax – or we'll fire you!** Keep things light. Nervous or stressed players can't expect to be creative. With happy, comfortable and relaxed players, the session will go smoother and all will inevitably play better.

Punching

• **Check, check mate.** Some sessions today don't use punch-ins. They record the piece a few times, then cut and paste to get the final result. For punch-ins, even though you're positive that you have the correct record-ready button pressed on your recorder, check again. One glance at the track sheet, and a double check of the tracks can save hours of repair. Listen to the track you are about to record into. Press the input button in and out and listen. Are you about to record on the correct track?

• **Use bars as your frame of reference.** Count the bars for each section. A 4/4 beat count is 1 2 3 4, 2 2 3 4, 3 2 3 4, 4 2 3 4, then back to 1. Maybe the tom-tom fill before a chorus starts on the 4, 2, which is exactly 2 beats ahead of the chorus down beat. Counting bars helps you punch in at the exact right beat. It beats using the 'I'll punch in on the chicka-chicka bit into the chorus' method. Write the number of bars per verse and chorus on the back of the track sheet.

• **Double check if you aren't sure.** Unless you are absolutely sure of the punch, go back and listen to the in and out spots. Players will be happy to wait. It is better to take five minutes double checking the punch than spending an hour repairing an erased part.

• **Make a safety before trying a difficult punch-in or punch-out.** Practice the punch using the rehearse button on the multitrack recorder. The machine will go into 'input' without recording, so you can get the feel of the in and out spots.

• **Turn the cue down.** Turn the cue level down before playing the music, especially if the player isn't expecting it. Many artists like the music in their headphones loud. When the song stops, of course the cue is quiet. Pressing 'play' half way through the song makes it blast out of the headphones.

Turn down the cues, warn him that 'here it comes' and press play. Raise the cue level in the headphones to the previous level.

• **Counter intelligence.** When punching in repeatedly at the same location, rewind to the same spot every time – about ten seconds before the punch. Enough time for the player to know where the 'in' spot is, yet not too long so as to waste time.

• **Don't punch in on the exact downbeat,** but a touch early, somewhere between the previous 'and' and the downbeat. Traditionally, this was done because the relays of the old two-inch multitrack recording machines took a couple of milliseconds to switch into record mode. Digitally, it works because it opens a little crossfade window a few milliseconds before the downbeat.

• **Be as specific as possible as to what you want the player to do.** Simply having him redo a track with no indication of what is needed helps no one. All singers need guidance from the control room. Granted this is not usually the engineer's job, but the producer's. Commonly, the more you engineer, the more you will learn about production.

• **Fill track sheets while recording, not before.** Things can change fast in the studio, and the track sheet must stay current. As well, double check that all information has been noted including specified song tempos, dates of recordings, players, song titles, digital formats, song structure, and any other pertinent information.

• **Watch those hands.** Get to a point where you can recognize chords, whether by ear or by watching the player's hands. It may not work so well with piano, but works great when a guitar player is right there next to you. You won't believe how impressed clients will be if you say 'Let's punch in on that second G chord' and then you do it.

Recording Vocals

• **Check with the singer.** Different singers will be prepared at different times of day. While one may be raring to go at 9 a.m., another may not really open up until the late afternoon. A tired singer sings a tired vocal. Schedule the vocal session for the singer's best time.

• **Encourage the singer to warm up,** preferably with scales in the key of the song to be worked on. No one can be expected to jump in on any creative endeavor without warming up first. Give him the needed time and privacy.

• **Get the lyrics.** Write the machine counter readout numbers on the lyric sheet to help locate the verses, choruses, and bridges. Going directly to the second chorus of the third take will be a breeze.

• **Encourage a singer to memorize the lyrics.** Something is lost when the singer is reading lyrics off a page. Better if they focus on the interpretation and feel of the vocal, not hunting around for the next line.

• **Hold that note.** If the singer is a guitar player and he is used to playing while he sings, by all means, give him a guitar when he does the vocal, even if he holds it without playing. Record the guitar too.

As well, if a singer is used to holding the microphone when he sings, let him hold the microphone. This creates a headache for the engineer, who wants to get the best recording possible. But better emotional perfection than technical perfection.

• **Wear a set of headphones during vocals.** When you wear headphones and monitor the cue mix, you hear exactly what the singer is hearing. This lets you fine tune the cue mix as the vocal progresses. Lower the control room monitor levels to avoid influence.

• **Kick everyone out of the control room.** Only essential people are allowed in the control room during vocals. Even the best of singers can find concentrating on vocal parts difficult with a room full of people staring at them.

• **Pump them up.** Get into it. Get the singer into it. It's much easier to record an inspired player. Keep the vibe up, be positive, and tell him when he does a good job. Tell him what you want, not what you don't want.

• **Let it flow.** Let the singer sing. Stopping and starting can be distracting to everyone involved. Let him run through the song to totally get into the flow of it. Invite him to listen back in the control room where you can work out any problems together.

• **Some days are just tough.** On those rare occasions when the emotions just aren't flowing, maybe tell the singer to picture one person in his mind. Forget the studio and the microphones, just picture that person, maybe an old girlfriend or a movie star, perhaps even a certain recording engineer, and sing directly to that person.

• **Tell the singer to sing along as soon as he hears the music.** This ensures he will have the same groove as the original, rather than starting cold on the downbeat of the intended punch-in. Once he knows where he is in the song, switch the track to input, so he hears himself singing. Punch in at the appropriate time.

• **'Why are they laughing at me?'** The singer can usually see the people in the control room, yet not hear them. If there is any delay at all in the control room, don't leave the singer hanging. Tell him exactly what is going on. Try as best you can to keep things rolling smooth as silk. Any delay may take its toll on the vibe of the vocal.

As well, if everyone in the studio bursts out laughing for any reason, even the most seasoned professional will wonder if everyone is laughing at him. Either assure him that it is not his vocal track causing the jubilation or, better yet, don't let this happen.

• **Bring the singer into the control room.** Leave the studio monitors on and do a vocal in the control room with the music blasting directly at the singer. This can be the best way to get a solid vocal. Set the microphone monitor signal just short of feedback with a non-omni-directional pattern.

Or he may want to wear headphones in the control room. If so, the engineer would wear headphones as well.

• **She's too flat, and not so sharp.** Singers, like everyone else in the world (including you), will have bad days. Some days they will sound absolutely magnificent, and some days they will sound like a train wreck. Your job is to tell whether the session time could be better spent on other things.

• **I hear that strain a comin'.** If the singer is pushing too hard, strained vocal chords come through loud and clear on the recording. And when a singer pushes too hard, the pitch often suffers. Check that the vocal level in the head-phone mix is loud enough for him to hear himself properly. Strain is not the same as growl or punch. Vocal strain usually means it's time to stop singing for today.

• **Record vocals every day.** Encourage the singer to record some vocals every day. All too often the vocals get recorded in a hurry on the last few days of the project, and all the work spent on great guitar and drum sounds will be wasted. Recording vocals every day gives you quite a few vocal passes from various sessions. Some will be average, but some may be outstanding.

• **Get a good lead vocal early in the project so the rest of the instruments can build around it.** The players need to hear that vocal track so they can stay out of the way. Plus a great vocal track early helps everyone else get inspired to do their best.

• **Have patience.** Any singer no matter how good, can lose that excitement after doing the same take hour after hour. If you lose patience with a player, it may not be long until he loses patience with you. The door swings both ways. Not everyone is a virtuoso.

• **Take a step back.** Once a vocal overdub is recorded, have the singer step back by a foot or two, then double the track with a matching vocal. Add this in with the choice vocal to give it more depth and placement.

• **Starting to feel thick.** For a thicker vocal track, try turning the live side of the microphone around, and have the singer do a track from the off-axis side. Add this track to the main vocal at a subtle level.

• **Double your pleasure.** When doubling vocals, ask the singer if he wants the live vocal in one side of the headphones, and the previous vocal pass in the other side of his headphones. This makes it easier to distinguish his live vocal from the recorded vocal.

• **Flaw hide.** Double or triple track vocals to mask minor tuning flaws.

• **Dim some.** Dim the lights and light the candles, burn the incense, take all your clothes off – create a mood to help the singer feel comfortable, relaxed and confident. The more at ease the situation, the better the outcome of the tracks. A strong vocal track makes the singer and you look good.

Pitch

• **Slow pitch.** If a singer is continually flat on certain parts of the vocal, use the VSO to lower the pitch of the recorded music. This may make it easier to hit the note. When he does hit the note, deactivate the VSO and return to the normal tempo. This slower speed during recording results in a higher pitch on the vocal sound when returned to the correct speed. Either record a whole track, then bounce the needed bits into the main vocal track, or just punch into the main vocal track at the altered speed. Of course, if he always sings sharp, do the opposite and raise the speed.

• **Cue down.** When a singer is having a hard time hitting notes, turn the cue mix level down, not up. If he insists on a loud mix, pull some of the lower frequencies. Loud lows can mess with a singer's pitch.

• **Take one side off.** To help him hear himself, encourage him to remove one side of the headphones so he can hear himself in one ear, and the cue mix in the other.

• **Record a reference track.** To help a singer follow the tuning and melody, record a simple piano or acoustic guitar track playing the vocal melody – no chords, just single notes of the melody of the vocal track. (Of course, not to be used in the final mix.) Add this track into the cue mix, and maybe remove any other instruments that may be throwing the singer off-pitch.

• **Don't use headphones, use speakers.** Sometimes using the opposite polarity speaker setup described on page 157 works to help the singer's pitch. With no cumbersome headphones in the way, the singer's pitch may improve.

Post Record

• **Stay in contact with the player.** Make all changes quietly and efficiently, without ever losing contact with the player in the studio. Change tracks when the singer is having a sip of water or looking at notes, definitely not when he is speaking to someone in the control room. Changing tracks from one to the next should be seamless.

• **Three ways.** If you are recording many passes of the same thing, either:

Use the individual track to monitor. Match the settings on all the intended tracks the same as the original. For example, if recording vocals on tracks 14, 15 and 16, match the cue levels, the send and return levels, and any equalization or processing on these three channels. When you change over from recording on track 12 to record onto track 13, it's set the same as the previous one – so the singer hears no change.

Use a single track to monitor. Patch the output of each track into a single master track. For example, you record the first vocal pass on track 13. When you change over to record another on track 14, rather than monitor channel 14, patch the output of channel 14 into the channel input of 13. All the subsequent tracks recorded are routed through channel 13, so the singer hears no change.

Use a single track to record. On a DAW, set up a master track and move each completed file down to a dedicated playback track. For example, you record the first vocal pass on track 13. Rather than record the next pass onto track 14, move pass 13 down to empty track 14, and record the next vocal pass on track 13. Set the 'lock' function to keep the pass 13 from shifting back and forth.

• **Minimize your choices.** Get the best sound, commit to it, record it, then move on. You will get much more done in the long run. Keeping multiple takes of different ideas to 'decide later' usually wastes too much time.

• **I've got more tracks – what can we record?** Just because you have extra tracks does not mean you are obliged to record on them. If the song sounds complete, it probably is. After a certain point, the more instruments that are added, the more each sounds a little smaller.

• **Problems?** Keep problems to yourself, and quietly work around them. Don't bring the session vibe down with your petty, 'Hey, I'm choking here.'

Recording Effects

• **Don't record the effects.** The traditional thought is, if the guitar effects are coming through the amplifier, record them. If the guitar effects are created in the control room, such as delay effects, don't record them. A fuzz pedal placed between the guitar and amplifier would, of course, be part of the sound coming out of the amplifier. You wouldn't record a clean guitar then add the fuzz later, as the fuzz sound will help the player play his part better.

If you plan on using the same sound during the mix, you might record the dry signal, then log the effects settings. This doesn't always work. All the proper paperwork in the world doesn't guarantee the same sound will return.

• **Record the effects.** Sometimes you find a perfect effect that helps the player become more creative, and musically, he goes places he normally wouldn't. Record the effects if they are part of the sound that influences the player. If you have available space, record the effects on separate tracks.

Punching in and out can be difficult when effects are involved, because punches can eliminate the decay of a reverb. Punches are smoother if no effects are recorded. Any little glitches and pops are smoothed over by adding reverb over the finished track

• **Don't go overboard with outboard.** You can always add more effects later, but you can't take an effect away once it has been recorded.

The Part

• **Assume that all tracks recorded will be keepers.** The first pass may be the best, even if it may not seem like it at the time. Sometimes it can be hard to tell while the recording is taking place, and only upon playback can you really hear what is recorded.

Keep a track if there is even a chance the player cannot redo it. How many great riffs have been erased, and replaced with a sterile copy of a great mistake? Too many to count.

Maybe tell everyone you want to hear the track one more time, and just quietly bounce the part to an open track, or set up another track to record on. If the player does the part better, great. If not, you still have the original intact.

• **Listen to the player.** Let the players hear the results. Discuss everything and consider their suggestions. Musicians know how they should sound, and what they want from their instruments. If you can take their vision, and bring it up a notch or two beyond their expectations, they might play a bit better.

• **Humble beginnings.** When a player is not up to a part, maybe suggest that he try an easier version of it first, then move up to the more difficult one. Once the easier version is recorded, if the harder one never gets done, you still have the easier one.

The age old way to get a good pass was to punch in the pieces until the best pass was recorded. Sometimes, once this pass is done, the player has an easier time playing along to the good pass.

• **It's not your job to hurry the players along.** Leave that to the producer. Do your work until they tell you to stop. Telling a player to hurry up has a negative effect.

Bouncing

• **Bouncing checks.** With limited space, combining two tracks to one track is common. When combining (called bouncing, or comping) two or more tracks onto one, listen back to the complete new comped track both by itself in solo, and within the song at various levels. The combined tracks may sound great together, but put them in the mix, and one of them may disappear. Erase the original tracks once the comp track has been approved.

• **To save tracks, combine tracks that don't overlap.** Different instruments may play at different times throughout the song. If, for example, a shaker is used in the chorus, and a tambourine is used in the verses, put them on the same track and split them later during the mix. This may require console automation, or an extra set of reliable hands.

• **When combining tracks, don't radically alter their sound.** Once they have been individually equalized, tracks bounced from many to few are unchangeable. Get a clean sound on everything then combine them. Trying to separate combined tracks is like trying to unscramble an omelette.

• **When combining tracks, radically alter their sound.** Change the equalization so opposite tracks are combined, such as a tuba and a triangle. When you mix, split the tracks and pull all the lows on one – the triangle – and pull all the highs on the other – the tuba.

• **I'd like to see your organ at ten o'clock.** When combining stereo tracks, pan them so that the center of each track is at a different spot in the panning perspective. For example, bouncing a stereo piano and stereo organ from four tracks to two, place the center of the organ at ten o'clock, and the piano at two o'clock.

• **Do not erase the count-off.** Bounce the count to a safe track. Without a proper count-off, there is no intro reference.

Doubling

• **Double a guitar track with different fingerings of the same chord.** Have the player use a different fingering to play the same chord. A capo, alternate string tunings or just different fingerings of a chord will make the double track sound a bit different, but still with the same chords and progressions.

• **Use the VSO when doubling.** Double a guitar track by slightly speeding the recorder up and recording a pass. Triple the guitar track by slightly slowing the recorder down and recording a pass. Pan these additional tracks left and right of the main track. The guitar sounds wide and chorus-like when played back at regular speed.

• **Double tracks must be tight.** Of course, a track sounds big by doubling it. But if double tracks are to be presented, panned and processed as one, sloppiness is not acceptable. Every bit must be doubled exactly, or the sound turns pretty vague pretty fast.

• **Metal rules.** Are you recording music with all electric guitars? Record a rich acoustic guitar doubling the electric tracks and bring it in just under the electric guitars. This can make a distorted electric guitar track sound more musical, full and rich. Keep the acoustic guitar level lower than the electric, or risk exposing it. Process it together with the electric guitar track so it fits in more as an enhancement rather than a specific acoustic guitar part with placement of its own.

• **Overdub more than one player at a time.** Musicians playing together will lock into a certain groove. Commonly, the result is better than if they played the same parts separately. I remember Keith Richards and Ron Wood recording their guitar overdubs at the same time. They both stood next to each other in the studio, stared each other down, and let it rip. They played off of each other in a way that wouldn't have worked had they each recorded their parts separately.

• **Record different guitars for different sections of a song,** such as guitar 1 in the verses, and guitar 2 in the choruses. This can add movement to the track as long as both are equal in power. If changing actual instruments is not possible, try different guitar pickups for different sections of the song.

• **I love your big bottom.** To widen the image of the bass, record a stereo track of a piano playing low octave bass notes. Add this into the mix for some stereo bass effect.

• **Listen back every time.** No matter how hurried the session, listen back completely to a final track before moving on. This may save hours in the future if an unnoticed mistake slips by during the recording. If a track needs to be repaired, do it now, rather than return to it another time. No engineer wants an unfinished track hanging over his head, and no artist wants a mix with half-finished tracks.

Edits

With the ease of digital editing, it is not uncommon for songs to be built from the best parts of any number of takes. Often the best sections are repeated, then nudged into time with the other tracks.

• **What are crossfades?** Digital editing consists of cutting and pasting files or regions of files together to create coherent sections. Crossfades ensure a smooth transition to seamlessly join these individual sections together. Crossfade parameters allow the user to choose between many curves, or even 'draw' a custom curve for a specific situation.

• **It's the music.** Like analog editing, digital editing is so much easier if you understand the musical technicalities of the song. Finding, for example, a specific four-bar section of a guitar track out of another pass means you must be able to hear the music, understand the changes and the timing, retrieve the part, and paste it back in at the correct spot.

• **Work on a copy.** With the ease of making digital safeties, keep the master intact and edit a safety master.

• **The urge to merge.** Before editing begins, use the pop-up menu to choose between the best parts of many takes. With each pass correctly labeled, the pop-up menu allows you to program different parts from different passes, as long as each part is properly labeled. This lets you hear the complete song with all the best parts from each pass before the first edit begins.

• **Don't edit the click track.** The click track is the song's timing reference, so editing it will render it useless.

• **Start editing the vocals from the middle of the song.** Often, as the singer gets into it, there are better choice bits to glean from. Find the best one, then tighten it up, then paste it into all the rest of the choruses.

• **HALT!** Consider the consequences of pasting the same vocal into every chorus or verse. Cutting and pasting a vocal track that someone has worked hard at can get tricky. Check with the singer before pulling all the choruses.

• **Don't solo when you edit.** Many edits you hear in solo are gone when the rest of the music is on. Don't spend hours cleaning up the tiniest detail that will be lost when the rest of the tracks are back in the mix. Other instruments in the monitor mix will help you hear which bits of the vocal are not quite in tune.

• **Don't get lost.** Use the digital editor to level out every word, so nothing – not even a syllable – is lost in the mix.

• **Nix the clicks.** Minimize clicks between edits by placing the reference lines where the waveform crosses the zero point. Zoom into the highest level where you can see the waveform.
     Sometimes where two sections butt up against each other, a 'click' is unavoidable. Mask the click by positioning it on a kick, snare or percussion hit.

• **A digital editor cannot create groove.** There is no substitution for a group of players watching each other and all being locket into a groove. This 'grease' simply cannot be manufactured. Tightening up the timing of a track will not automatically improve it. If the track is so out of time, maybe redo it. Or maybe that's the player's style.

End of Session

• **Print a rough mix.** After finishing a major track, print a quick rough mix for future reference. Mixes from different days throughout the project may have different feels and different grooves.
     Maybe have the assistant write all the settings down for each rough mix you do. If one of the roughs has a certain feel that everyone likes, use the assistant's notes from that mix to refer to on the final mix.

• **The future is limitless.** Sometimes after a long day of recording, your equalization and compression levels have crept a bit too high, due to ear fatigue. First thing tomorrow, listen to the sound with fresh ears and check for overprocessing. Next time maybe back it off a bit, relying on experience over tired ears.

• **Remove all digital tapes before turning off the recording machine.** With 'rotating drum' tape machines, the digital tape wraps around a drum inside the machine, so turning the machine off and on can stretch the tape. See Digital Appendix for more on rotating drum technology.

• **Rewind all tapes before putting them away.** Normal the console, help clean up the studio, and return any borrowed gear from other rooms on the premises. Return any wiring changes, such as inputs to a machine, back to their original connections.

• **Store all the microphone accessories in the same place.** Keep all extra clamps, windscreens, external pads, etc. in the same place for easy access when making a quick change or addition. Store the microphone shockmounts with the microphones so they won't get lost.

Some studios leave the microphones set up all the time, with cables on them ready to be used. Over time, this may be a detriment to microphones, as a diaphragm can react to moisture, dust, smoke, and changes in temperature.

• **For safeties' sake.** Make safety copies of everything and print out a final track sheet to keep with the recordings.

• **Go ahead and back up.** Back up data as a safety precaution. Although many times backups are not needed, it is that once-in-a-blue-moon where data is lost. Whatever the format or situation, if there is money or talent on the line, back up your digital data.

• **Count everything.** Count the microphones at the end of every session. If something goes missing, ultimately the recording engineer is responsible.

• **The music stays with us long after the client has gone.** No tapes or hard drives leave the studio until cleared by management.

• **Stay until the end.** Don't leave the session until all the rooms are cleaned, every piece of gear is put away, all documentation is complete, and the place is ready for the next session.

Or not. If nothing is to be changed, run a strip of tape across the console so no one touches it, and tape no entry signs on all the doors.

• **My studio is a total write-off.** Use daily work orders – even if you use your studio alone. Keep track of every day in the studio, what happened, when you started, when you finished. When the daily log is signed every day, there will never be any question about hours used.

• **Clean the patch cords.** Scrubbing off the buildup with brass cleaner will eliminate crackles, and will keep the runners in the studio busy. Have them clean the cymbals and the hardware on the drums too.

• **How does a clean guitar sound?** Clean the 1/4" guitar jack inputs with a bit of contact cleaner and a nylon-wire bristle rifle cleaning tool, available at any sporting goods store.

• **Erasing all the tracks.** Clean the console of any pencil marks, tape or general studio muck.

• **It's a clean machine.** Over time, dirty console knobs may occasionally crackle with use. Commercial spray cleaners are available to keep the knobs and dials on the console clean. If you make a slight equalization change while recording, any crackles caused by the dirty knob get recorded as well.

• **I've got some great contacts in the studio.** Spray your cable connectors and contacts with commercial electronic spray gunk cleaner. Better contacts result in better sounds. Do a before and after listening test and see.

• **Tired of your grimy old console?** Take a day to remove all the knobs, put them in a laundry bag and do a cold water wash. Air dry them and they will be clean as new. Take the opportunity to clean the console too.

Chapter Eleven

THE MIXING

Writing on mixing is a difficult task. Try explaining to someone, without actually being there, how to paint a picture, how to play the blues, or how to remove a spleen.

This chapter is just a list of basic starting points. There are, of course, a million different variations on everything mentioned here, and the enclosed just scratch the surface.

• **The trick is, there is no trick.** A good mix is akin to a jigsaw puzzle, where there is a perfect space for every piece, and all spaces are covered. The image should have width and depth.

The best mixes come from well-written, well-arranged, well-played and well-recorded songs.

Pre Mix

• **Leave some time between the recording and the mixing.** If possible, don't record overdubs and mix on the same day. A good mix normally takes at least a full day, and recording something will throw off the flow of the mix.

When you must do an overdub before the mix, use the input button on the recording machine to verify the correct signal flow. As tracks are often cross-patched, or patched into different channel line inputs, it would be a shame to record over the kick drum while intending to repair a vocal part.

• **Clean up your room.** Mix in a non-cluttered environment. Music sounds better in a tidy room the same way your car runs better after being washed.

• **Normal the console.** It can be frustrating to unmute a channel to find the line trim has been turned up to maximum from the previous session. Switch the console to the proper operating status.

• **Set unity gain.** Use the oscillator to set unity gain on your compressors, limiters, delays and some equalizers. Use a pink noise generator to set levels on the reverb and effects units.

• **Wire those speakers there.** Use the best professional nearfield monitors available to you, along with highest caliber speaker cable connected to a studio standard amplifier. Check that the speakers are wired properly.

Use familiar speakers. Setting the right low end can be tricky until you know exactly what your speakers and control room are capable of.

• **A/B CDs.** If you are not familiar with the mix room, play a favorite CD mixed by a top gun engineer against a project that you recently mixed. Are the high, midrange and low frequencies all there? A/B the two tracks to get a sense of the room and speaker response.

If, for example, your mixes come across bass heavy, maybe the low end in your mix room is not properly equalized. The same principle applies to midrange and high frequencies as well. Tuned or not tuned, the size, shape and absorption factors all contribute to the final sound. Once you understand the room, you can judge your own mixes.

• **Is the song 'mix ready?'** Starting fresh is always better than wasting time cleaning up tracks. Before starting the final mix, confirm that:

–   All choice tracks are returning properly and correctly labeled.

–   All recording and mix machines are cleaned and aligned. All clock rates, parameters and plug-ins have been checked.

–   All files have been correctly loaded and checked.

–   All vocals are in tune and in time.

–   All music tracks are clean, in time, and in tune.

–   All drum tracks, including samples, are just as you want them.

• **Check your session notes.** Review all the notes on ideas that seemed to work while tracking and overdubbing.

• **Listen to the rough mixes.** Sometimes a ten minute mix at the end of the session can sound better than spending ten hours on it come mix time. Rough mixes may be heavier, lighter, drier, etc. Listen to which direction captures the feeling best, and aim the boat that way.

• **Mix with your heart, not your head.** A certain mindset is needed to start fresh on the mix. Paint a mental picture of how you want the mix to sound, then start with a fresh, relaxed attitude.

And just because you aren't supposed to do something is no reason not to do it. Something may not be what you are used to, but if it seems to work in a track, go along with it. Keep it fun and you get better results.

• **Don't start with the best song.** When mixing any project of more than a few songs, you will find there are one or two 'best' ones. Don't mix these best songs first. Wait until you are knee deep in mixing. Hit your stride, then mix the best songs.

It's like making pancakes. Get the bugs out of the mix with the first few, then proceed.

Setting up to Mix

• **Buss transfer.** If applicable, transfer all tracks to the least amount of tapes. Machines need time to lock when they switch from rewind to play, and this lockup adds up to wasted time. Two machines lock faster than three.

• **Pattern the tracks the same during all your mixes.** Maybe return the drums on the first twelve channels, then the bass tracks on the next channels, then guitars, keyboards and vocals. Get used to having the same instruments return on the same channels throughout the mixing.

• **Return the lead vocal to the middle of the console.** Set your vocal sound with your head right between the speakers, not way over to one side.

• **Azure mixing or purple pro's.** Use different colors for the various sections on the scribble strip of tape across the console. Pink, blue, and green page highlighters can represent drums, guitars or vocals. After mixes the scribble strip is kept with the mix information. If a mix needs to be recalled, the scribble strip is an essential part.

• **Check the track sheet.** As tracks are returned to the console, mark the part off on the track sheet. When every musical part on the sheet is checked, all tracks are returning properly. You don't want to mix a song, then realize you missed a track.

• **Assign the kick drum and bass guitar to a subgroup.** The interplay between the kick and the bass is crucial. If one is changed, the other may sound off. When the mix needs a touch more low end, preserve this interplay by moving the subgroup, rather than the individual faders.

   Subgrouping instruments together early, such as all the guitars, or all the background vocals, may help you when experimenting with different level changes, pans, mutes, and ideas.

• **Fix it before you mix it.** Go through all the tracks and erase any unwanted noises such as coughs or instrument sounds. If you have a totally automated console, maybe just mute the bad bits. Realize that this is the time to be extra careful about not erasing anything. If automation is not available for the mix, make a digital safety track and practice the cuts, then fix the unwanted bits.

• **Eliminate the mistakes.** If a section of a track is musically questionable, options include:

– Lowering the level at the worst parts, then raising the level of another track to mask it. If this doesn't work, then mute the track, as long as the client doesn't object.

– Finding a similar section of the song, and sending it to a sampler, or 'cut and paste' it to replace the questionable section.

– Having the player re-record the part, but this probably means setting up the amplifiers again and hoping to match the sounds.

– Leaving it out. If the flaw happens early enough in the song, maybe leave the track out until the start of a bridge or chorus. The best way is to record the track correct the first time around.

• **Activate the loop mode.** The uniform repetition of the song playing over and over can help with the flow of the mix. Often, like the musicians playing the song, the mixer can get into a 'creative groove' as well.

• **Set up all your needed effects.** All outboard effects should be patched in and checked, with the scribble strip across the console properly labeled. Use a small strip of white tape to label each effect with the appropriate send and return.

Maybe patch all the effects for the instrument with the instrument on the console. For example, if the drums return on channels 1 to 9, then return the drum effects to channels 10 to 15.

Perhaps recall a favorite snapshot as a starting point.

Mixing Basics

• **Spending twenty hours on a mix will not make it twice as good as spending ten hours on a mix.** At some point, the best has been done and continuing is fruitless. Put it to bed and go home.

• **Bring the best things forward in the mix.** The so-called 'hook' of a song is its defining entity, or the central core of what keeps it going. Identify one or two fundamental elements and accentuate them. Enhance the groove and preserve the emotion and feel.

• **Build on the best properties.** Build the song like a staircase – step by step. Introduce instruments as the song progresses rather than giving the listener everything at once. Create movement throughout the mix by changing the levels, panning, effects and processing.

• **Solo you can't hear it.** As with the recording process, don't press solo too often. It's great to use the solo button to get a basic sense of an instrument, or to find a problem, but get in the habit of changing equalization with the rest of the tracks in the monitor mix. When you can't hear the other tracks, you can't effectively equalize a track to fit in.

• **Don't spend too long on any single instrument.** If you are working on, for example, the drums, get a basic drum sound then move on and tweak it through the mix. Don't spent six hours getting a perfect drum sound, because everything changes when you bring in the rest of the instruments.

• **Relax your ears.** Don't mix for hours upon hours without taking a silence break. Your ears need time to relax and rejuvenate. Your ears are organs, not muscles – overuse does not make them stronger. If that were the case, I would have a liver of steel.

• **Forget the clock.** Don't hurry the mix. Be creative and get it right the first time. Expect the mix to take longer than you expect.

• **Run the faders at their optimum level.** Pushing faders all the way up adds distortion, so set them around zero, their optimum operating level, then fine tune the levels with the gain trim. This will send appropriate levels to any pre-fader effects.

For clearer, more transparent mixes, set the gain trims as low as possible, and set the master buss level at zero. This is crucial on budget consoles when distortion increases as gains are boosted.

With the master fader always set at zero, you know if it has been moved or not, and you know where to return after every fade.

• **Don't go upstairs.** Don't mix using the biggest studio monitors unless you are totally familiar with them. Think of the average listener. Using the biggest honking speakers on the planet may make the mixes sound great in the studio, but most listeners use average size speakers. Make your mixes sound good on smaller speakers so you know that they will sound good on the average car radio or TV speaker.

• **Dissociate yourself with the mix.** Many times you will fall into the trap of thinking, 'What will other engineers think?' or 'Will the radio play this if I do some radical idea?' Liberate yourself from any preconceived ideas as to what the mix should sound like.

• **If a part doesn't fit, don't try to make it fit.** Mute it and be done with it. Each element must be solid. If the song has three guitars playing roughly the same thing, either clean them up, or use only one of the tracks as a main one. In the long run, each remaining instrument will have more impact.

• **Don't eliminate parts without authorization.** The mixer's job is to mix, and the producer's job is to tell the mixer what to mix. Don't take it upon yourself to pull musical parts unless you are the producer. The last thing you want is a client calling you a week after the mix asking where his favorite keyboard part is.

Processing

Equalizing

• **It's there so I must use it.** Properly recorded tracks shouldn't need drastic changes in equalization. Overequalization contributes to distortion and vague sounds, and creates peaks at certain frequencies. In a good mix, all relevant frequencies are audible. In a bad mix, all you hear are the equalization peaks.

Equalizers, compressors and noise gates are tools, and should be used for repair. See Chapter Eight to help you determine your individual instrument's equalization and compression settings.

• **Consider where will the sound lie in reference to other sounds.** Placement of the instrument will influence how it is equalized. If, for example, a non-feature track is hard panned to the left, you may not want to equalize it the same as you would if it were a feature track that was panned toward the center. Non-feature tracks should have certain frequencies pulled to increase the clarity of the feature tracks.

• **Don't add every frequency to every track.** Pick and choose where to add, where to pull and where to leave alone. Pulling, or notching frequencies on some instruments opens windows for others. If you have one instrument covering all the low frequencies, you might need to pull some of its frequencies to make space for something else. Determine which tracks take up which frequency areas and where any overlaps occur.

• **Keep it musical.** Use equalization to enhance the musical tones and downplay the less musical tones. See the chart on page 115 to determine where the musical tones lie.

• **Use equalization to curb excess noise.** Turn down the high frequencies on a noisy bass amplifier or a noisy reverb return, then run the track through a harmonic distortion device. This may enhance the low-frequency harmonics, and bring back the higher harmonics with less noise.

• **Depth to infidels.** Use equalization, compression, reverb and panning to place instruments sonically in front, behind, above or beside another. A less prominent instrument can appear to be placed in the background by lowering certain frequencies on the track and its effects.

For example, if the lead vocal track has a boost at 3 kHz, maybe pull that frequency on the background vocal tracks to create a 'bed' for the lead vocal. Maybe add a touch of reverb to place it more in the background.

• **Pull one frequency to make the other shine.** Find and pull a low frequency with a thin Q setting, then boost an octave harmonic of that frequency to bring out more of the musical aspects while maintaining a clean solid sound. This works well on the bass guitar tracks.

• **Is equalization really improving the sound?** As you change the equalization settings, periodically A/B between the unprocessed sound and the processed sound to see if your changes are an improvement.

• **Use level changes rather than EQ changes.** With two similar tracks in the mix, pull the higher frequencies from one track and the lower frequencies from the other. Combine these two to get a proper blend of highs and lows. Rather than adding high frequencies with the equalizer, simply raise the level of the brighter track.

• **So low the bass and kick.** Pull unwanted low, lows on some (but not all) of the heavier tracks. Determine which tracks, often heavy guitar, bass guitar or kick drum, are adding unwanted vague rumbles. Rumble is not to be confused with valid low frequencies of the kick drum or the bass guitar.

Listen to where your changes start cutting into the impact of the track, then back off a bit.

• **If it sounds good, do it.** Sometimes you have to turn a frequency way up, to heck with what all the rules say. And if it sticks out too much in the mix, the mastering engineer will ease it off.

Compressing

• **There is no wrong or right compression.** Some tracks might need lots of compression and some tracks might need none. Some engineers prefer to use as little compression as possible, while other engineers compress everything. Compression settings should depend on the needs of the song, not the habits of the recording engineer.

• **Compress the vocal.** Even though the vocal was probably compressed during the recording process, it may need additional compression in the mixing process. Because it is the most important factor, it must stand out. To avoid doubling up on the same settings, use a different compressor when mixing than was used on the tracking session. Note that some recording engineers never 'double compress' a vocal.

The compression settings used while recording are a rough estimate of the perfect settings, because during the recording process, no one knows what the final outcome of the song will be.

Once all instruments are recorded, then the engineer can hear exactly what compression settings are needed. For example, if all the rest of the instruments are heavily compressed, some of the vocal lines may get lost if the vocal isn't processed similarly.

• **Use the best compressor, limiter, de-esser, equalizer and console strip** on the primary instrument, usually the main vocal. Some consoles may have new or updated strips with better processing circuits installed. Use these strips for the vocals. If the studio has only one really great equalizer, use it on the main vocal track to expose the 'air.'

• **Keep all levels consistent throughout the song.** Odd notes here and there must be ironed out, whether that means using compression, riding the levels or even bouncing the offending track to another track with appropriate moves. With consistent levels, you can ride a track knowing exactly how it will react.

• **Place a de-esser across the send to the vocal reverb.** High sound pressure levels on 'S' and 'T' can create a spike that may easily overload the input of the effect.

• **Instant punch.** Good kick drum sounds have the power to rattle your ribcage. That solid punch comes directly from the movement of the speaker. The over-compressed kick drum loses that punch because it can't move the speaker the same way. Set the threshold higher rather than lower, use a high ratio, then raise the level of the kick drum track.

• **Yes to de-ess.** Use a de-esser in the snare drum track to minimize high hat leakage. A de-esser also works to pull the 'screech' or string noise from an acoustic guitar track.

• **Use the local buss system.** Blocks of tracks, such as drums, electric guitars, and vocals, can benefit from additional overall compression. Use two busses to send the drums into a stereo compressor. Return the compressor back to two channels on the console. Assign these two channels to the main output buss. Set the compressor to a high ratio, fast input, and medium output until you get a real tight hard sound. Raise the level until it adds enough punch to the drums. Go over the top for a real trashy sound.

• **Free samples.** Often, if you don't want to use a snare sample, or if one or two of the hits in a song don't ring quite right, sample a good hit from the same song, and add it in with the rest of the snare. Then you aren't really changing the sound of the drums, only enhancing it with the same drum sound. Some artists are sticklers about this. They want no 'foreign' material, or samples from other sources.

• **Chains required.** If compression isn't quite enough to make the vocal stand out, run the vocal into the sidechain of intrusive instruments, perhaps a piano or lead guitar, maybe even the whole mix. The vocal dictates to the compression, so the other tracks are automatically lowered when the vocal is in. Of course, when the singing stops, the other tracks return.

Gates and sidechain

• **Open the gates.** Insert noise gates on the room tracks and set them to open whenever the drummer hits the snare drum. He hits the drum, and the gates on the room tracks open for a large ambient snare sound, then close again. This keeps the kick drum out of the room sound and gives the snare added size.

• **Close the gates.** You might get a great snare and tom-tom sound, then bring up the overhead microphones, and there goes the great snare sound. All you hear is wash. Insert noise gates in duck mode on the overheads and set them to close whenever the snare drum hits. This lets the original snare track come through without the influence of the overheads.

Note that this means your cymbals may not come through as expected.

• **Clean up a track.** Set a group of instruments to trigger off a main track so all the tracks open and close uniformly. This can vastly clean up sloppy tails, when players don't start or end a phrase together. This works great with, among others, bass, background vocals, horns and keyboards.

Maybe use the kick drum to trigger a bass that is always ahead. Insert the noise gate into the bass channel, and set a very fast attack and a slow release. Send the output of the kick drum to the sidechain input of the gate, so the bass track opens when the kick hits. Mess with the settings until the kick drum and bass guitar react as one.

• **Use the oscillator to add low end to the kick drum.** Sweep the oscillator to find a nice low frequency, preferably in the key of the song being worked on. Send the oscillator through a noise gate, then trigger the input off the existing kick. Whenever the kick hits, the gate opens and this nice low tone is added to the kick drum sound.

Traditionally, recording engineers have used loud oscillator levels to effectively expel annoying record executives from the control room.

• **Pink or white, I just can't decide.** Use a pink- or white-noise generator to help with the snare sound. Run the white noise through the noise gate, and trigger it off of the snare drum. The gate opens then closes with the snare, adding fullness to the snare sound.

• **Setting on the 'duck' at the bay.** The vocal is usually the most important thing in the mix, so clarity is paramount. Try running the main reverb through a noise gate on the 'duck' mode. He sings and the reverb ducks to give the vocal some room. When he stops, the gate opens and all the reverb returns. This also works great on rhythm guitar as the reverb is only really heard at the tail end of the musical phrase. This tricks the listener into thinking the reverb is on throughout the musical passage.

• **Buenos de-ess.** To make a vocal cut through, higher frequencies are added and that increases increasing sibilance. A good de-esser will control sibilance problems while maintaining the high frequencies. Use a de-esser on any high pressure spike, such as vocal sibilance, string noise, even high hat leakage.

• **Chop shop.** Use the high hat or click track to trigger that choppy pulsing sound effect. Great on synthesizer tracks.

(1) Use a delay to double the existing quarter-note click track to an eighth-note click track.

(2) Set the noise gate at a very fast attack and release.

(3) Insert the noise gate into the synthesizer track.

(4) Run the output of the delay into the sidechain input of the noise gate.

(5) Mess with it to get the appropriate 'chop' sound.

• **We have a pulse.** For a good pulsing effect, set the auto-panner at extreme left and right. Set the timing to the beat of the song, then monitor one side of the stereo auto-panner from full on to quiet. This pulsing effect works well on, for example, a keyboard pad. Maybe trigger the timing off the kick, so every pulse happens in time.

Effects

Reverb

• **What is reverb?** Figure 11.1 shows reverb as a series of dense multiple reflections of varied repeats that are blended into a smooth decaying tail. Reverb in today's studio usually means a digital effects unit or program. Different companies have different products, but most effects units follow the same pattern. All have presets that are easily changeable to suit every situation. Typical parameters may include:

Halls: Good for all purpose reverb or echo effect. Parameters allow users to create any size or makeup of space.

Rooms: Used more for ambiance than reverb. Great for drums and other percussion instruments. Can give life to dead sounding instruments.

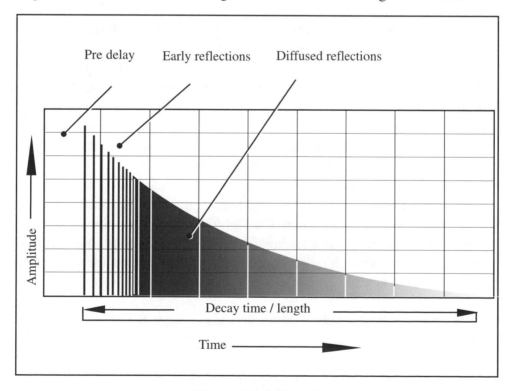

Figure 11.1 Reverb

Plates: Vocals and acoustic instruments sound large, lush and thick when sent through a plate. Before digital reverbs, studios used reverb plates – a metal foil steel plate with a speaker device on one end, and a pickup on the other. Signal caused the plate to naturally vibrate. The pickup sent the signal back to the console. A moveable damping pad allowed for different lengths of reverb. Better studios still have a reverb plate or two.

Size: Perceived size of the room that the reverb simulates. When set for a larger hall, the signal would last longer than a small room.

Decay time: The length of time the effect takes to return to 'silence.'

Pre-delay: The delay between the send and its effect. In real life, a sound takes time to reach an opposing surface then return, creating a natural pre-delay, or echo. The human ear is used to this effect, so natural sounding pre-delay enhances depth perception.

Early reflections: The single reflections that occur before the diffuse, reverberant reflections occur.

Diffusion: The denseness of a sound. Denser being thicker, darker and more complex.

Low-/high-frequency roll-offs: Some units have internal roll-offs, and allow the user to roll off low or high frequencies.

Crossover frequencies and levels: Some units have crossover settings that allow the user to set where the high/low crossover point is, and how much low or high frequencies are desired within the sound. You may, for example, want the high frequencies of the reverb to last a bit longer than the low frequencies.

Input and output levels: I think you know what the input and output controls do.

Mix or wet/dry: This determines how much of the original signal you want returning with the sound. Works great when recording, for example, an in-line guitar track where the signal is routed directly through the unit and then recorded.

• **What is the difference between echo, reverb and ambiance?** An echo is specific repeats. Reverb tends to be shorter, more dense repeats and ambiance is referred to as the shortest of reverbs. You hear echo when you yell in the Grand Canyon. You here reverb when you yell into the Grand Hotel ballroom. You hear ambiance when you yell into the Grand Hotel washroom.

• **Don't put every reverb on every track** or every track will be lost in the mud. Make a sound larger by adding something dry for comparison. In a wetter mix, something dry will stand out. In a dryer mix, something wet will stand out. Effects used sparsely will leave more room for all instruments to cut through.

• **Don't send everything to one reverb.** Send minimal tracks to any one effect. Reverbs are used for placement and depth for each instrument. Sending too much of everything to one effect defeats that placement.

But if two tracks, for example a vocal and a guitar, never play at the same time, you might use the same effects on both of them as the effects would never overlap.

• **Send everything to one reverb.** Sending, for example, all the drums to the same effect, at proper levels, can place drums in a specific room or locale, rather than having one effect on the snare, and another on the kick, and yet another on the tom-toms.

• **Mixing the manufacturers gives more variance.** All of the companies that manufacture effects units create their own programs and algorithms for their effects. All equipment the manufacturer releases uses these algorithms.

Using two effects units from the same manufacturer on, for example, a vocal track, means two sends are going to the same basic algorithms. When sending the vocal to more than one effects unit, all should have different manufacturers.

Cheaper effects units are not always true stereo. The left side is just a phase reversed right side. Listen in mono and the effect disappears. Defeat this by using two similar units in tandem. Split the input and send it into both units, and listen to the left outputs of both units as your stereo returns.

• **Don't wash your bottom.** Use reverbs sparingly on the low-end instruments such as bass guitar. Sounds with lots of bottom can add unwanted wash when sent through a long reverb unit.

• **Use a reverb to mask minor flaws,** such as an abrupt stop or a punch. Camouflage the blips by using a reverb setting that has the same sound quality as the track. Try setting the same equalization on the track and the reverb unit so they both become one, sharing a common space in the mix. Determine what you want before going in and messing with the internal reverb settings.

• **Use the reverb plate for the vocals.** If you are lucky enough to have a real reverb plate, use it for vocals and principal instruments in the mix. Reverb plates and spring reverbs are great to place an instrument or vocal exactly where you want it. Warning: they may be noisy.

• **How long is a pre-delay?** Set the pre-delay at anywhere from 40 ms on up to a few hundreds of milliseconds, depending on where it feels best. The pre-delay is often set at a multiple of the beats per minute of the song. The longer the delay is, the less it becomes a pre-delay, and more of an effect itself. Any pre-delay less than 40 ms is considered part of the sound itself.

• **Give the vocal its own reverb.** When the vocal track has its own reverb, you can make all the changes and fine tuning you want without affecting any other instruments or effects.

• **Use two reverbs on the vocal.** Sometimes two reverbs combine to get one great reverb sound. A short reverb then another longer reverb. A longer reverb with less low end and a longer pre-delay brings out the shorter brighter one.

• **Stereo returns from effects don't necessarily have to remain stereo.** Use one side of an effect for specific placement of an instrument, panning the instrument and the effect the same. Or pan the instrument to the left, then pan the effect return to the right. Maybe pull the high end on the distant side of the reverb.

• **Oldie but a goodie.** Send a dry signal through a guitar amplifier in the studio. Use a send from the console, and set it to work with the amplifier input level. Place close microphones in front of the speaker to get more of an electric sound, or place microphones in the distance, using the room as a live chamber. For ambiance, place two microphones in opposite corners, about a foot out and aimed into the corners.

• **Delay left, right?** To get a real feeling of depth from a live room, place one microphone in a corner of the live room, and the other one in a nearby hallway. Due to microphone placement, signal will hit one microphone before the other. This creates a slight delay on one side of the audio spectrum. Pan the instrument toward the side with the early hit. In the track, different delays create a real sense of width. Check for phase issues.

• **Send one effect into another.** There is a world of sounds available just by sending one effect into another, such as reverb effect into a chorus, or chorus into a delay, or delay into a reverb. The combinations are endless.

• **Check your reverb returns in mono.** Why? Because, even in this age of modern videos, many televisions have one little speaker, and out-of-phase mono signals will cancel each other.

• **Pedal to the metal.** Bring in effects pedals, delay pedals, distortion pedals, chorus pedals and all the gear used traditionally for live guitars. Overload them, distort them, dunk them in the sink, try any combination for creative sounds.

• **Pull certain frequencies on a send, then add the same frequencies on the return.** Theoretically, they should cancel each other out, but you may end up with some interesting sounds.

• **Equalize the send to an effect before equalizing the return.** Pulling an errant frequency from a track before it reaches the effect leaves the effect return unprocessed.

• **Lower lows.** The faster the tempo, the less low end is needed on reverbs.

• **Faster songs benefit from shorter reverbs.** Leave the long reverbs for the slower songs. You lose all the effect after a certain tempo.

• **Set the end point of the reverb to be complete by the time the next beat hits**, especially on percussion instruments where long reverbs can overlap into the next beat.

• **Long reverbs can clutter important space.** Sometimes an equivalent delay can create the illusion a long reverb without the additional cloudiness. Keep longer reverbs lower in level.

• **Place tracks in the distance using a less dense setting.** In nature, a sound source loses density with distance. For closeness, try increased density.

In a painting, a light shaded background pushes the darker shades foreword. Similarly, a dense setting on one effect can push a track with a less dense setting into the background.

• **Set the crossover frequency at the key of the song.** To set the crossover point, go into the internal parameters of the effects unit. The crossover points are lower in level. This may help open up a frequency area that's already full.

• **Rather than equalize a track, maybe equalize the effect instead.** Adding a bit of high end on the effect may be enough to bring a track out of the background. Of course, this will affect all tracks being sent to that effect.

• **Equalize the output of digital reverbs.** Change the settings at the unit, rather than at the console. Why introduce another processor (more noise) into the chain at the console?

• **Mud wrestling.** Be aware of the track equalization being sent to the chorus or harmonizer effects. For example, the equalization on a bass guitar is +3 dB at 100 kHz. Sending all those low frequencies into a harmonizer will return as harmonized mud.

Split the signal into two channels. On the second channel, pull the lower frequencies, then send that signal to the harmonizer. Of course, once you get your sound, switch this second track out of the main mix buss. The returns of the effects will have richer bass frequencies with minimal muddiness.

Delay and Chorus

• **Delay lay delay.** Delay is the interval between a sound and its repetition. Echo is a series of repeats of an original sound that gets duller and quieter on each repeat. A delay unit in the studio simulates this allowing the user to set the length, amount of delay, modulation, and many other settings. Delays can be modified to work not only as a standard delay, but as a chorus, a flanger or a doubling effect. Figure 11.2 shows that standard delay units contain:

Feedback regeneration: The amount of signal fed back into itself. A signal with no feedback will delay the original signal once. As the feedback increases, the delay repeats itself as it gets quieter. Set too high, and each repeat returns a bit louder than the previous one, creating feedback.

Delay time: The distance in time between the original signal and the first repeat of that signal.

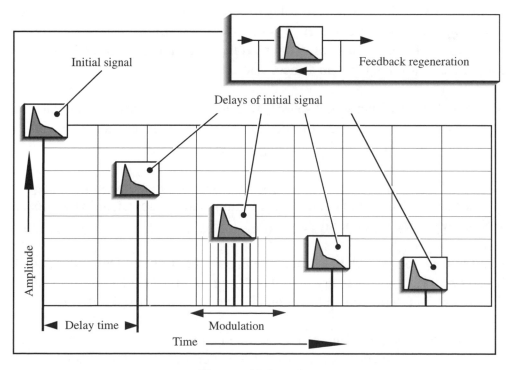

Figure 11.2 Delay

Depth, or modulation: This function allows the user to determine the movement of the delay. For example, the delay can be set to shift from 100 ms to 120 ms, resulting in a less rigid sound.

Speed or VSO: The speed of the modulation sweeping from the highest and lowest settings on the modulation. It can be slow, up to a few seconds, or fast, as fast as a few milliseconds.

• **What are chorus and flanging?** Chorus and flanging are functions of delay, with options that also include depth, modulation and speed. Chorus is an effect in which a signal is combined with its pitch-detuned replica.

Flanging is an effect where a signal is combined with its delayed replica. The delay is so short (0–20 ms) that the direct and the delayed signals cannot be resolved – they act like a single sound. Varying the delay creates phase cancellations that move up and down the frequency spectrum. Chorus and flanging really must be heard to be understood.

• **Thickly sweet.** Return a stereo delay to separate channels on the console, then slightly delay both left and right sides by a few milliseconds. Try prime numbers. Different settings of depth modulation and delay times can result in flanging or harmonizing. Pan as necessary.

• **Whirled series.** Rich sound sources with lots of harmonics sound great with a chorus effect. Use chorus to add thickness to acoustic guitars or vocals. Engineers differ in viewpoint on using chorusing on a piano.

• **Wide you do that.** On slower songs, chorus effect on a bass guitar can give a wider effect. But too much big thick bottom can eat up valuable headroom, which lowers the level of all the tracks.

• **Extra fat.** Record a double track to enhance the original. Delay the double by 20 or 30 ms and set a varying delay. Bring it in under the main instrument for fatness and depth. Keep the depth and volume low or risk pitch issues.

• **Change for a single.** Run the bass through a pitch change program set at an octave higher, or lower – yikes. Bring it in under the primary bass track.

• **Use a chorus effect to mask slight pitch problems.** Sending a slightly out-of-tune vocal through a chorus effect can mask pitch problems. Sending a really out-of-tune vocal through the chorus effect will accentuate pitch problems.

Sometimes doubling or tripling a vocal track can mask tuning problems, causing the whole vocal to sound somewhat in tune.

• **Pitch it.** With auto-pitch devices available, no final vocal track should have tuning issues. But get the best vocal you can to avoid the auto-tuner, then only tune what is necessary. Too much auto-tuning can pull all the emotional nuances out of a vocal.

• **I'm down.** Harmonize the drum room tracks down to get a thicker, heavier 80's rock drum sound.

• **Flange like they used to.** The traditional way to flange a sound was to send a track to a second analog recording machine. Using the distance between the record head and the playback head on both machines, they would press their thumb against the moving reel, or flange, on the second machine. The original sound meshing with the returning track from the second machine produces the flange effect.

Another old way of getting a natural flange was to place a microphone aimed at the cone as normal. Have someone hold a second microphone by the cable. As the microphone hangs, they slowly spin it around. As the signal goes though the speaker, add the spinning microphone in with the first microphone, then mix them to hear the effect.

• **Swish on the verse, swoosh on the chorus.** Sometimes you might want the chorus or flanger to swoosh a certain way every time. A slower modulation setting dictates where it randomly swooshes so the effect changes with every pass. To get the same exact every time, record the swoosh onto two tracks a few times until it is exactly right. Maybe help it by changing the modulation settings at the appropriate times, or changing the delay times.

Practice it a few times recording each one until the recorded effect sounds correct. Once recorded, the effect occurs the same every time. It uses two tracks, but it frees up the effect for other instruments.

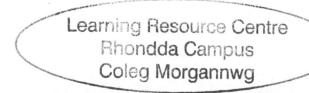

• **Dig it all delay.** When a digital drum signal is sent to a digital reverb input, the reverb reacts the same every time. Insert a delay between the send and the input of the digital reverb. Set the delay time short, with no feedback, and adjust the VSO for medium speed and low depth. With a slight varying pitch input, every hit into the digital effect will be slightly different in timing and pitch.

• **Delay the ambiance tracks to make the room sound larger.** Either use the same delay setting on both left and right side of the stereo room tracks, or go kooky and set the delays at different times.

• **To get a nice thick guitar sound,** buss the signal to a pair of delays, both with short but different delay times – under 100 ms. Add a bit of modulation on each delay then pan them hard left and right. The guitar is occupying that middle space normally lost during hard panning.

• **Use a very short delay.** To raise a track's high frequencies without adding equalization, send the track through a quality delay set at less than 40 ms. Pull the low frequencies, maybe add some compression, then bring this second track in under the original track.

• **Split with an effect.** Use a chorus effect to split one track into two, such as changing a mono keyboard pad to stereo.

• **What is slapback.** Slap, or slapback is an ancient method of analog delay used by our engineering grandfathers. These pioneers created a feedback loop, taking advantage of the distance between the record head and the playback head on vintage analog recording machines.

In those days, the term 'most significant bits' was in reference to the studio receptionist.

• **Drum slap.** For a longer snare drum sound set the delay at 1/8 or 1/16 note, then send the snare drum to it. Add some feedback to the delay and bring it in under the original snare drum.

• **Delay of the land.** For a nice stereo spread effect, set one delay to match the 1/4 note beat of the song, and set a second delay at 2/3 of the beat. Work with the delay times, panning and feedback until it suits you.

Strategically set delays can create great spaces, and be used for chorusing, looping, reverb, pre-delay, mock stereo, and slap.

• **Delay sounds.** For more distinction between the vocal track and a delay, set different equalization settings on the delay than on the vocal track. For less distinction between the vocal and the delay, set the equalization to match.

• **Halo, I love you.** To get that halo effect around the vocal, set a 1/4 note delay with the high end pulled a bit. Slowly bring the delay in until you hear it working, then lower it a touch. Turn the overall mix way down, and listen for the delay. Try adding reverb and experiment as desired. It should be felt rather than heard. Personally, I'd like to be felt rather than heard.

• **Distance yourself from the band.** Rather than using the feedback on the delay unit, create feedback by sending the signal from itself through a channel with high frequencies lowered so each delay consecutively lowers in fidelity. This can make the delay sound natural, but not so the listener says 'Hey, that vocal is going through an effect.'

• **Experiencing long delays.** Long delays work great as an effect, but can add clutter if always left in the mix. Maybe clutter is the desired effect.

• **Delayed due to snare.** If a single snare shot needs to be replaced, delay the hit ahead of it by the exact time distance between snares. Then record the new snare, actually the delayed previous one, over the faulty one. This will require a delay and an extra channel, often not a luxury while mixing.

Set proper levels and always double check when you record into a snare track. Maybe record a safety track first.

• **Use two alternate samples for one snare sound.** If you must redo all the snare hits, don't use one sample throughout. Set up a panning system with two samplers. Find two similar snare hits and sample both. Set the panner to trigger with each hit. When one snare hits, the panner is triggered over to the other side, then sends the signal to a second snare sample, after this one hits it goes back to the first sample. This really works well during drum rolls, where a single sampled snare might be exposed.

• **Missed her tambourine, man.** To place a tambourine hit on every other snare drum hit, use the same 'panner into samplers' method as above. With a tambourine hit in one sampler, and silence from the other, every other snare hit triggers the tambourine hit.

• **Less is louder.** All the equalization, reverbs, echoes, and choruses add level to the master buss. Too much of everything means the whole mix lowers in level. Less processing takes up less space so it naturally sounds louder and better.

• **Low on effects?** So what? You don't necessarily need racks of outboard to do a good mix. One or two reverbs, a delay, maybe a chorus can get great results.

• **Too many effects.** With the glut of digital effects and mixing gear available today, it can be easy to overdo it. If an effect helps the flow of the song, that's great, but to add some kooky effect just because you can contributes to unneeded clutter. Your job is to clear away debris, not add to the mess.

Levels

• **Listen to where each track lies in relationship to the others.** Bring in all the tracks and set some basic levels and pans. Commonly the kick and snare drums, bass guitar and lead vocals are in the middle, with the rest of the instruments strategically panned.

Some engineers start with the drums and bass as they are the foundation of a song. Some engineers start with the vocals because that is the essence of the song. But every song is different, so use the first few passes to listen to all the tracks, then determine your best approach.

• **Pull the faders down and start over.** Sometimes it's hard to get started. If you lose the groove in the levels, just pull down all the faders and start over. Bring each instrument back into the mix and set your new levels.

• **Bring the vocal in early.** All the instruments will eventually surround the lead vocal. Bring the lead vocal into the mix early to hear what changes are needed in other tracks to keep the vocal sounding good.

• **Turn down not up.** Before changing one track's level, try lowering another track, to make the first track jump out a bit. Raising tracks because they keep getting lost means there may be a equalization problems. Check your settings to see which frequencies are overlapping.

• **A slight change in panning can coax more clarity out of a track.** Before changing a track's level, check if a simple change in panning works. When two instruments with the same frequency range are similarly panned, they mesh together. Pan them apart for distinction.

• **Keep things from jumping out in the mix.** Small level changes might work better than drastic level changes. Mixing isn't about perfect levels, its about getting a feel across to the listener. It's an art, not a science.

• **Spotlight.** Aim a spotlight on the vocal and, when the vocal is not on, aim the spotlight on a guitar solo or a drum fill. Between vocal lines, push the guitar so the spotlight is on it, pull it down when the vocal returns. Focus on one thing at a time within the mix.

• **Let me roll it.** For emphasis on snare or tom-tom rolls, boost only the drum reverbs, not the drums themselves. This may give them enough push without actually changing the drum level.

• **Keep the vocal right up front in the mix.** In today's popular music, the most important element of the mix is the vocal. Make it sound as rich as possible with every single note heard.

Placing the vocal in the mix correctly has a lot to do with the singer. Some vocal tracks just sound loud no matter where you place them. A strong vocal performance might not need to be placed as prominent in the mix, as it will naturally project loud and clear.

Just because you have a great sounding guitar doesn't mean it always has to be loud. Guitar lines, drum fills or anything else shouldn't mask the vocal.

• **Left, right?** Position the panning so tracks are even throughout the song. Not necessarily balanced at the same time but throughout the song. For example, placing a cowbell on the left side during the choruses and a tambourine on the right side during the verses creates a balance in terms of the total song.

• **When panning, check in mono.** Phase cancellations may cause tracks that are loud and clear in stereo to vanish in mono. Change your panning while in mono to make some instruments disappear and others emerge from the mist.

• **You're right, my instrument left.** Set stereo pan-pots around nine o'clock and three o'clock, rather than hard panning. Hard panning stereo tracks such as drum overheads or a piano can remove a crucial middle space. As a rule, lower frequency instruments are less directional, and get panned more toward the middle. Sometime bass and kick are not panned exactly at twelve o'clock, but panned off a bit, so they aren't sonically right on top of each other.

Many engineers love hard panning. Old Beach Boys, Beatles, even Alice Cooper recordings contain lead vocals or drum tracks panned all the way to one side. It was a favorite trick to add dimension to a song.

• **Split that guitar in half.** Try splitting a guitar track into two channels. then reverse the polarity on one of the two channels. Can be kooky. Check in mono because the signals will cancel out in mono if set at the same level.

• **Six dBs of separation.** Stereo acoustic guitars sound wonderful as a featured instrument, but stereo instruments take up a lot of space. As more instruments are added to the mix, the less the stereo image remains.

• **Divide and conquer.** Split one channel into two channels for flexibility when sending a signal to an effect or through a compressor. Use the second channel to change the reaction of the equipment without changing the sound in the monitors.

The high and low frequencies of a track may need different compression settings. A properly equalized instrument should hit the compressor equally, keeping all levels intact.

Try splitting the return from the vocal into three channels on the console, and process each a bit differently for different sections of the song.

• **Mute me on the high C's.** Try different ideas with the mute buttons, such as coming in half way through the time signature. Come in on the 2, and out on the 4, just to hear what happens. Muting one track leaves space for another track to shine.

• **Use channel mutes to send specific hits.** When you want specific reverbs, such as on a single snare hit or a reverb on a certain word within a line, use a send, not a return.

Return the track to a separate channel, turn up the reverb send, and remove the channel from the main stereo buss. Unmute the channel on the specific hit.

• **Speed it up.** If a mix doesn't seem to have a lot of life, sometimes speeding up the tempo of the machine a notch or two will add some spark. However, some artists don't like this. I worked on a Rolling Stones record where the producer tried a tempo change as an idea on one song. Keith was not impressed, claiming no one messes with Charlie's grooves.

• **Placement in the mix is crucial.** Create depth and dimension within the mix by placing instruments close and far. Place drier brighter instruments up front, and duller ambient instruments in the distance. Aurally place a track in the distance by:

— Lowering the volume. Louder instruments will appear closer, so lowering the level can lessen that closeness.

– Panning toward one side. There is a one in 360 degree chance a distant sound originates in front of us. Instruments, such as vocals, panned to the center provide equal energy to both left and right speakers making the instrument appear close.

– Easing off the high end. When a sound originates in the distance, the high frequencies dissipate before low ones. When you hear a neighbor's stereo, you don't hear the crispness of the track, you hear the low end thud. A track without high frequencies can sound distant.

– Adding effects. When a sound originates in the distance, we also hear the wash of the natural echoes and reverbs.

• **They panned my mix.** Place the instrument using reverb panning. For example, place an instrument at two o'clock, and the panning of the reverb at eleven o'clock and five o'clock.

• **Plug your ears.** While mixing, you might lose a bit of objectivity. Try this. Set the volume at a reasonable level. Plug your ears with your fingers, close your eyes and listen. This seems to give a different perspective of levels, and is a good method of checking the vocal and snare drum levels.

• **Catch a buzz.** Listen to the mix through headphones to catch any buzzes, clicks, pops, hums, etc. Tiny flaws sometimes not evident in the monitors can come through loud and clear in the headphones.

Listen through headphones at a low volume for a true feeling of instrument placement and level. A lot of listeners enjoy their music through headphones.

• **Don't run out of power.** If you run out of CPU power because of the amount of plug-ins, process (record) some of the files that have compressors, limiters, de-essers or equalizers, then remove these plug-ins. This will free up processing power because playing back a track requires far less CPU power than playing back a plug-in in real time.

• **Smile.** Bring in a camera, and let the clients take photos of the sessions. Those little throw away cameras work great, and they keep everyone occupied while you fix that piano track you erased.

Print Prep

• **The stage is late – take the buss.** If you use overall compression across the master buss, wait until the latter stages of the mix to activate it. As more instruments are added during the mix, the compressor changes the sounds you have already tweaked.

• **Two passes to master.** Set up two mix machines and print one mix with no compression and print another mix with ample compression. Let the mastering engineer choose which he wants to use.

• **It's time you converted.** If you are mixing to digital format, use only the highest quality A to D converters available, even if you must rent them.

• **Print alignment tones.** Before printing your mix to digital tape, print alignment tones at the head. This gives the mastering engineer the proper references to master. With a standard level of 0 VU on the console, set the input level of digital machines to -12 dBFS to -20 dBFS, depending on studio practice.

Some engineers feel these tones are not needed. While these tones are used traditionally on analog machines to adjust operating level, azimuth, bias, and high and low frequencies, these parameters are not user adjustable on digital machines. This debate continues.

Record a tone on the left side first, then both sides. The mastering engineer reads this information on a label, then confirms that the left and right sides haven't been accidentally switched.

• **Mix to two.** Mixing to two tracks digitally is a poor man's automation. Record the mix to two available tracks of the digital multitrack tape. Because the mix occurs exactly at the same time as the tracks are played, punching in and out should be seamless. For example, a rhythm guitar is a bit low in the second verse. Raise it a bit then record, or 'punch in' to the mix for the second verse. Maybe, if space is available, print different passes on separate sets of tracks and build one great mix.

• **Mix to analog.** Many engineers agree, and for good reason, that mixing to analog sounds better than mixing to digital. If you get the opportunity, mix a project to digital and to an analog mix machine, then A/B the result of your mixing experiment. You can't tout the great sound your digital recorder gets until you can compare it with analog.

• **DAT's new to me.** Your mixes are important. Use fresh tapes, not ones that have been previously recorded.

• **Exercise your tapes.** Rewind any virgin mix tapes to the end, then back again. This gets the tape and the machine 'used' to each other. On analog tape, this is referred to as exercising.

• **Is everything clean and ready?** All your equipment must be reliable and maintained to a professional standard. Check that the mix machines have been cleaned, aligned, loaded with formatted tapes and drives (if applicable) and are ready for use.

• **Run a safety of all the mixes.** DATs, CDs, and hard drives are inexpensive, Safeties are usually always usable if, for some reason, the final mixes become lost or unusable.

• **See these CDs?** If the final mixes will be printed on CD, set the DAT mix machine at a sampling rate of 44.1 kHz – the sampling rate of CDs. The signal will not need to be converted from a different sampling rate.

• **Add some 'life.'** To polish up a final mix and give it a bit of life, add some overall equalization and compression. Done properly, this can add some life to them, especially if the mixes aren't going to be mastered professionally.

–   Set unity gain on both the best equalizer and compressor available.

–   Connect a stereo equalizer to the main buss of the console, keeping the L/R sides correct.

–   Run the stereo equalizers into a stereo compressor, or vice versa, then back to the stereo buss insert return.

- Start with a high threshold and small ratio, and change them as necessary. A slower attack time may allow some punch to come through before it kicks in. For example, the attack on the drums can pull down the rest of the instruments. A slower attack time would miss the punch of the drums, and react to the less transient instruments. Longer release times will decrease the pumping and breathing effect. Unless the mix is routed through a multi-band compressor, the lower frequencies will dominate how the higher frequencies react.

- Less meter movement means fewer audible changes. This could mean too much compression.

- Don't hit the overall mix buss too hard. Turn down the overall buss limiter, and use compression on the individual tracks to allow the mix to breathe a bit more. Too much compression inhibits the speaker movement. This means the volume knob may be set louder, but the punch is still missing.

- Are you doing this process after mixing for twelve hours? Maybe wait a few days until your ears are fresh before compressing the mix. Compare your compressed mix with the uncompressed mix, and then compare your mix with some commercially released projects.

Printing

• **Before printing to stereo.** Ask a few questions, including:

- Is the low end punchy and solid, with both the kick and the bass guitar holding their own space?

- Are the midrange frequencies clear and full?

- Are the highs crisp, clean and clear?

- Is each instrument properly panned?

– Is each instrument correctly placed sonically in front of or behind other instruments? All tracks should sound clear individually, yet fit in with all the rest of the tracks.

– Do the drums sound like a solid cohesive drum kit, or are too many effects taking away from the groove?

– Do I know all my moves, including all level changes, cuts, pans, special moves and the final fade?

– Are all levels correct compared with other levels. Specifically is the snare drum too loud? Turn the master volume down and listen to the level of the snare drum to verify the proper level.

– Does the mix have liquid movement, with builds and releases, ebbs and flows and interesting changes? Movement in the mix is about bringing things in and out, up and down, back and forth. That motion helps the music from getting boring.

– Does the vocal stand out, with every syllable clear and audible?

– Is any link in any of the signal flow chains overloading?

– Do all sounds combine to make a single cohesive, solid, interesting piece of music?

– Did that sushi taste funny to anyone else?

• **Print hot.** As with recording the basic and overdub tracks, print mixes as hot as possible to digital. Hotter levels mean the signal is higher into the 'most significant bits' area. More bits used mean the recording is fuller, better. Take full advantage of the usable space between noise and overload.

• **Bring in a few extra hands to mix.** Delegate one section of the console per person, and let them do different things during the mix. Often a mix tends to be a one person job, with others telling them to please try this or that, especially with today's studio automation. When everyone is at the console trying different things as the mix is being printed, there will be many different ideas recorded for edits in different parts of the song.

• **The prints of mixes.** With today's digital technology, printing lots of mixes allows you to get exactly what you want. Print mixes with vocals up, vocals way up, vocals down, bass up, bass down, dry, wet, a cappella, it's endless. Take the time to listen and determine which parts of which mix are best. Do the edits with your notes close by.

• **Print a mix with no lead vocals.** Be sure to turn off the effects on the vocal as well. Previously this was done so the artist could go on TV and have the music behind him, then do a live vocal. Today it is more for safety's sake, in case something on the vocal track needs to be changed after all the mixes are done. This saves having to do a complete remix just to fix one part. The mix can also be used to edit out any naughty words for radio airplay.

• **Print a mix with lots of lead vocals.** Listen to some old records and hear that some vocals are far louder than the rest of the instruments. Print a few passes with various vocal levels in case a line in the master pass needs fixing. A louder vocal track may keep the song moving if it's a good vocal. When the vocal is too loud, it can diminish the impact of the rest of the tracks.

• **Undermix.** Lower the effects, maybe remove a track or two. Different emotions tend to emerge from different levels in the mix. Don't underestimate the power of a simple, dry mix of a well-written, well-recorded song.

• **Overmix.** After everyone agrees that the final mix is done, print an 'over the top' mix, going a little crazy with the sends and returns – but not so much that the pass is unusable.

Take it over the top, whatever you interpret that to be. What comes across as over the top in the studio comes across as dynamics to the listener. Like movies that have evolved from boring still camera scenes to explosions and movement on every scene, listeners today want more than a standard song, they want fast paced changes. You owe it to yourself, the client and the whole recording industry to go all the way to outer space if necessary when trying new ideas. No one wants the same old stuff. Wouldn't it be great to be known as a cutting edge recording engineer?

• **Let someone else drive the boat.** When listening back, let others sit in the engineer's chair. This makes everyone feel more involved in the project. Who doesn't like to sit at a mixing console and hear themselves play? Turn it up so everyone can enjoy it, then retreat to the hallway to rest your ears while everyone enjoys the mix.

After Printing

• **Listen to the mix through all speakers,** including small speakers, mono speakers, speakers in another room and car speakers. Today's portable 'boom boxes' are a great reference.

Using various references gives a more rounded and overall landscape from which to listen. If, for example, the vocal isn't loud enough on every set of speakers, maybe you should use the mix with the louder vocals. Listen and compare. If it sounds good on all systems large and small, you have a good mix.

Note that in the car, the background noise of the vehicle will tell you if all sounds in the mix can be heard. If any sounds are masked by the vehicle noise, make the sounds more apparent.

People love to listen to music in their car. Maybe turn up the bass and the treble on the car stereo to check how that sounds.

• **Make safety copies of the master mixes.** There is no reason not to have abundant safety masters. Make copies, or clones, of all the mix data. Pop out the safety tabs on all recorded tapes, especially on masters. On DAT tapes, it is a little sliding lever that, when open, makes recording impossible.

• **Proper documentation is paramount.** Label everything, including all song titles, band name, producer, engineer, assistant engineer, date, song choice and location, and all other relevant information. Tracks and sections of tracks are commonly shuffled not only from track to track, but from format to format. Clearly mark all the choice passes and locations. Without proper labeling you may wade through dozens of passes before finding the pass you want.

• **Listen up.** If your small studio does not have big honking monitors, take the mix to a bigger studio and ask if you can listen on the big speakers.

• **Leave the mix until tomorrow.** Leave the mix set up overnight so you can return in the morning and do a couple of tweaks then print a final few passes. But too much twiddling at this stage can undo all of yesterday's hard work.

• **Stacked or what?** Mix all of your various versions, master, vocal up, vocal down, solo up, T.V. and instrumental to a new session. Add enough tracks to this session so that all the mixes can be stacked in time, e.g. 1-2 master mix, 3-4 vocal up, 5-6 vocal down. This way you will be able to quickly edit a new 'master mix' comp by editing across all the mixes at once. This new mix will now be 'in time' immediately, with no guess work in involved.

• **Consolidate crossfades.** Improper crossfades can result in noticeable 'clicks' at each of these points. Consolidate the file to create a new continuous file, because the fewer edits a track has, the more easily the hard drive is able to find it and play it back. All hard drives have a finite amount of data it can 'through-put' at once.

• **Watch the levels.** If you are assembling master tapes yourself, use your ears not the meters to match the different levels. Depending on the peak content of a song, the meters may differ from song to song.

Use the meters as reference, but your ears as the final judge.

Mastering

• **What is mastering?** 'Mastering' is where the mixes are given the final equalization, compression and (sometimes) assembly. A good mastering engineer will match the sound and levels of all the songs, as well as bringing a transparent 'radio' quality to the project. He will have the proper listening space and gear. His trained ear from years of experience will bring the final mix up to professional standards, sort of like the icing on the cake. Some mastering houses will forward properly coded masters to the pressing plant.

• **What about Bob?** Get the best mastering engineer available. You will never regret making a record sound a little bit better.

• **Why, master, why?** Send the master tapes, or drives, with all the alternate mixes and out-takes, along with all proper paperwork and locations clearly labeled, to the mastering lab. The mastering engineer needs access to alternate mixes. For example, if he feels one section of a song needs louder vocals, he can retrieve whatever he needs from a properly labeled alternate take.

Sending the uncompressed originals is better because clones and copies are subject to human error. More transfers mean more chance of lost data. As well, if you recorded two versions, one with compression and one without compression, let the mastering engineer hear your compressed versions and get his view on your skills as a mastering engineer.

• **Listen to the master.** Listen to the changes the mastering engineer is doing, and ask questions. Compare and analyze the original mixes with new changes. This is the final stage. Don't judge the final result at the mastering lab. Take it around and listen for a few days. Any minor changes are easily communicated to the mastering engineer.

Chapter Twelve

THE DIGITAL APPENDIX

Digital recording systems store and retrieve numbers. Big numbers. These systems are broken down into three stages. Input processing, storage, and output processing. Output processing is input processing in reverse, so if you understand the input process, you get the output process.

In its simplest form, analog sound is a linear series of individual waveforms. Digital processors change the linear aspect of the analog signal to specific steps of varying voltages. Not unlike a cartoon, where what appears to be continuous motion is actually a series of maybe 16 drawings per second, stored in a specific order. As each drawing takes 1/16th of a second in time, the cartoon's sampling rate would be 16 (if cartoons had sampling rates.) Similarly, each 'sample' is recorded as a specific voltage for a specific length of time.

Each voltage sample is read and assigned an equivalent number. With sampling rates of up to 48,000 times a second and higher, these are some pretty hefty numbers. How does the computer store and retrieve them so fast? By converting the numbers from our familiar decimal counting system to a simpler binary counting system.

Digital Conversion

Binary codes

Do you have a social security number? I have. Do you have a driver's license number? I have. Do you have a 'cheese of the month club' member number? I have. One thing can be represented many different ways.

The eggs in Figure 12.1 are commonly represented by the decimal number 12 – '1' block of ten, with '2' left over. In an octal system (8), these eggs are represented by the number 14 – '1' block of eight with '4' left over. In a binary system, they are represented by the number 1100.

The binary system is far easier for a computer to deal with, because the computer recognizes 0101, or 0110 far faster than 9 or 4. There is little room for error when there are only 2 choices, 1 or 0, on or off. The actual numbers themselves are longer, but they still represent the same amount. If human beings had been born with one arm and five fingers instead of two arms and ten fingers, we would probably count using a pentamel (5) system rather than the decimal (10) system. Plus way more people would be named Lefty.

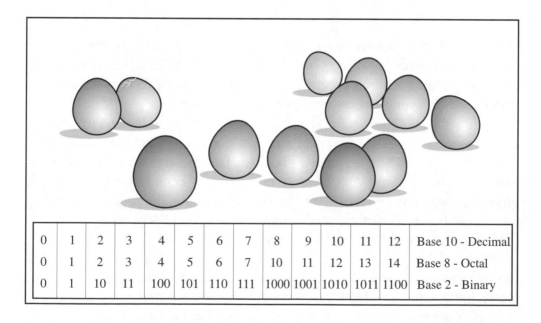

0	1	2	3	4	5	6	7	8	9	10	11	12	Base 10 - Decimal
0	1	2	3	4	5	6	7	10	11	12	13	14	Base 8 - Octal
0	1	10	11	100	101	110	111	1000	1001	1010	1011	1100	Base 2 - Binary

Figure 12.1 Different counting systems

Bit processors

The processor stores numbers in bits (BInary digiTS). To count to 12 using the binary system, 4 bits are needed (Figure 12.2). Switch that statement around, and a 4-bit processor holds up to 16 decimal numbers. For the system to hold more numbers, or finite steps, a larger processor is needed. A 16-bit processor can hold up to 65,536 numbers, and a 24-bit processor can hold over 16,777,000 numbers. Longer bit length means more finite steps, and a more accurate A to D conversion.

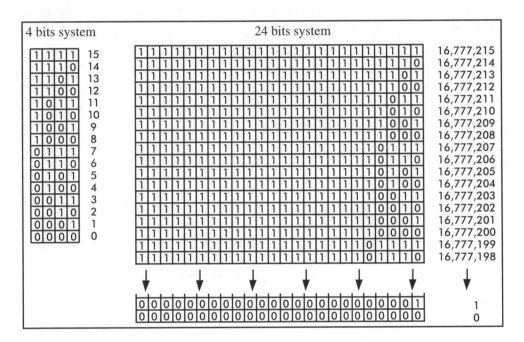

Figure 12.2 Low bits vs. high

LSB vs. MSB

The lowest digital levels – all zeros – are referred to as the LSB, or Least Significant Bits. The highest levels – all ones – are MSB, or Most Significant Bits. Low incoming levels activate the least significant bits, and hotter levels activate the most significant bits. The idea is to take advantage of the most bits possible. Lower levels mean fewer bits are being used, resulting in a 'grainy' sound, plus increased noise when the level is raised.

Input Processing

Dither

Before the signal can be converted to digital, the input signal needs to be processed. When an incoming voltage level is not strong enough to reach the lowest quantization level, digital distortion is introduced. Dither is low-level noise (white noise, square wave, or sawtooth, depending on the equipment manufacturer) added to the signal to mask this distortion by increasing the level beyond the lowest quantization threshold. This moves the incoming audio level up into more significant bits, producing a more accurate conversion of analog signal. Not unlike bias used in analog recording, this additional signal is added to improve reproduction.

Adding dither is recommended when converting a signal to a lower bit depth such as converting a signal from 24-bit to 16-bit. Dithering retains some of the 24-bit quality in the 16-bit signal.

Aliasing

Aliasing refers to errant frequencies introduced into the audio spectrum. The Nyquist theorem states that aliasing will occur when the sampling frequency is not at least two times the size of the audio frequency. So what happens when

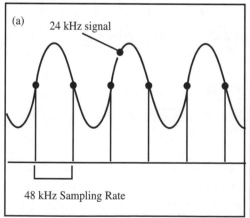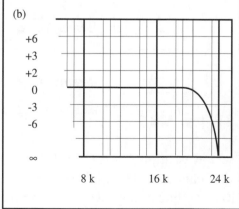

Figure 12.3 *Anti-aliasing filter*

the sampling rate is set at 48 kHz and the musical overtones reach 24 kHz and beyond?

Figure 12.3(a) shows that when the frequency equals or is higher than half the sampling rate the signal is no longer accurately reproduced. Figure 12.3(b) shows how a low-pass filter, also called an anti-aliasing filter, passes the low frequencies and stops any frequencies that may cause aliasing.

In this case anything under 24 kHz can pass through. To sample at a rate of 88.2 kHz, half would be 44.1 kHz. The anti-alias filter would remove all frequencies over 44.1 kHz. A sampling rate of 96 kHz would need an anti-alias filter over 48 kHz.

Sample and Hold

The A to D (analog to digital) conversion is done mainly in two steps, first, sample and hold, then quantize. With a 48 kHz sampling rate, each voltage step is held for a duration of 1/48,000 of a second, assigned a corresponding binary number (word) then stored and stacked as packets of information (Figure 12.4).

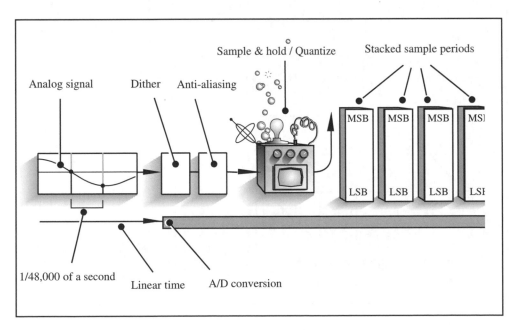

Figure 12.4 A to D conversion

Quantization/error correction

If a binary number is not available for the exact incoming voltage, the converters round out to the closest available number. Figure 12.5(a) shows a 4-bit processor with an 11 kHz sampling rate. If an incoming voltage level is 0.46012 mV the reader would assign it to the closest available numbers, maybe 0.5 mV.

Figure 12.5(b) shows a 16-bit processor and 48 kHz sampling rate. Much less rounding out is applied, vastly lowering the quantization error. An incoming voltage of 0.46012 mV would be recorded as 0.46012 mV.

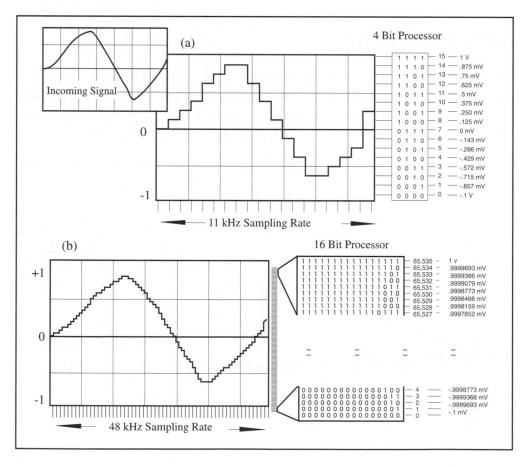

Figure 12.5 *Quantization/error correction*

Oversampling

To preserve as much of the higher frequency content before the anti-aliasing filters, oversampling widens the frequency bandwidth and allows for a smoother and extended frequency slope. Figure 12.6 shows the process that averages out the distance between two samples, adding another two samples (or four samples or eight or on and on). This doubles (or quadruples) the sample rate. This linear quantization reduces phase distortion and lowers the noise floor.

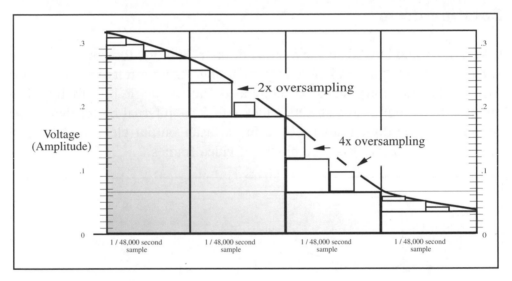

Figure 12.6 Oversampling

Digital Storage

Pulse code modulation

It's one thing to convert numbers to binary, and yet another to physically store them for future reference. PCM, pulse code modulation, is the actual switching mechanism between a zero and a one. The data stream switches from zeros to ones using voltage pulses that modulate between on and off. The medium records a pulse of voltage as a 1, and no pulse as a 0, all within a specific timed framework referred to as the clock rate. Like an orchestra leader the clock keeps all operations synchronized to a common reference time.

• **What is jitter?** Jitter is the slight timing variations of the sampling period. Ideally, a track should be recorded and played back at a perfectly fixed clock rate, but in the real world, minor fluctuations occur. These fluctuations can skew the waveforms on playback, causing jitter. Jitter is a function of the A/D, D/A conversion, and does not occur when the signal remains digital.

Digital Multitrack Recording

Rotating drum

DAT, ADAT and DA-88 all use rotating drum technology (borrowed from the video world) to achieve high-quality recordings using a relatively slow tape speed. The incoming signal is combined into a single-bit stream, then inputted to the buffer at a specific clock rate. This internal buffer then 'time compresses' the digital stream by using a faster output clock, resulting in time-compressed even and odd blocks of video frames.

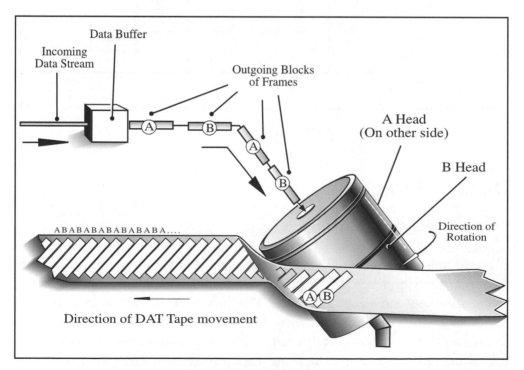

Figure 12.7 Rotating drum

Figure 12.7 shows how this system uses two heads, A and B, on opposite sides of an angled rotating drum. The angle reduces distance between video tracks, and increases surface area for data. The digital tape is pulled in and wrapped one quarter of the way around the drum. As the drum spins, the tape slowly moves in the same direction. Each time the head and the tape meet, the head lays down a burst of signal from the data buffer.

Stationary heads

DASH (Digital Audio Stationary Head) is an open reel digital tape format used in professional studios. These are very stable machines, available from two tracks to 48 tracks. The heads are stationary and not rotating, so the tape speed must be much faster. An open reel system, rather than video cassettes are used. Advantages of stationary head recorders are:

– As yet, no HI-8 (DA-88) or VHS (ADAT) cassette can hold over 8 audio tracks of professional quality audio, although multiple 8-track recorders can be linked. Some DASH machines hold up to 48 tracks of digital audio on a half inch thick tape.

– Some allow the user access to the digital buffer to be used as a sampler. This allows the engineer to fly bits of music around.

– The ability to (carefully) do razor blade edits – impossible with rotating drum technology.

– The elimination of wasted time waiting for machines to synchronize every time the engineer presses play. Often projects come into a big studio, and the digital program recorded from their ADAT or DA-88 is transferred over to DASH 48 track, or even to a hard drive for mixing.

Hard drive

Magnetic tapes are classed as linear storage because they must be rewound or fast forwarded to locate specific spots. Hard drives are classed as non-linear storage, or random access. The most common form of non-linear storage would be the standard CD. Information is suspended on the disc, and as the disc spins within the drive, the roving head has almost instant access to anything on the disc.

The hard drive disc controller spins a sealed non-removable stack of recordable magnetic discs. Between each disc is a roving head that both reads and writes digital information. Modern hard drives contain upto thirty-two heads for sixteen stacked discs rotating at 10,000 RPM. Discs move independently of each other, and are always in motion. The user cannot access the specific discs as the whole system acts as a seamless storage.

Files

With non-linear storage, a complete three minute song would normally be classed as a single computer file, accessed by clicking the icon. This complete file would be then loaded into the buffer to be accessed.

The hard drive is broken up into volumes, also called partitions. These volumes hold blocks of files. Each volume is independent of others and each has a directory indicating file size, type and location within the volume. Sound files are best stored in adjacent blocks, but sometimes it's impossible. Fragmentation occurs when smaller files within a volume have been erased, forcing the heads to break up a large file into smaller files to store them on these smaller spaces. This can really slow the system.

Today, with removable hard drives, a complete session can easily be taken from studio to studio. And no need to download. You can just work off the hard drive.

Compact disc

The compact disc is today's standard of pre-recorded consumer discs. Today's compact disc is used primarily to store a stereo audio signal. Increasingly, CDs also carry additional information, including computer games, programs, live action video, dialog – all forms of multimedia – stored within the same binary stream. At present the industry standard for CDs is 44.1 kHz 16-bit, so no matter what sampling rate or bit depth is used for the recording, eventually before the final release, it will be transferred down (or up) to 44.1 kHz 16-bit.

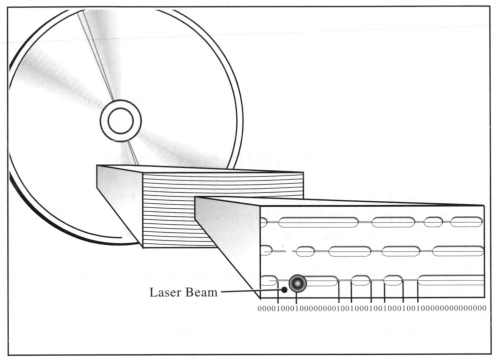

Laser Beam

00001000100000000010010001001000100100000000000000

Figure 12.8 Pits and flats on a CD

While the rotating head converts the stream of frames to time condensed blocks, the CD records a single stream of CIRC encoded data. The speed of the CD revolution changes as the disc plays, ensuring the data is read at a consistent rate. As well, the data is stored from the inside of the disc going outward, rather than starting at the outside, as with vinyl records.

Figure 12.8 shows a continual stream of different length bits and flats embedded on the plastic of the disc. Within these bits and flats is the binary data. One would think that the pits would be stored as 0's and the flats as 1's. Not so. When the pits change to flats, and the flats change to pits is where the binary data is stored. The laser records the change from either pit to flat, or back as a '1'. The rest are '0's.

DVD

The DVD is essentially the same size as a standard CD, but the standard CD holds up to 650 megabytes of stereo digital data, and DVD can optimally store up to 17 gigabytes. These discs are able to store far more data of high-quality video plus full bandwidth audio. How does it do this? By employing several different methods including:

– A higher pit density rate for tighter tracks and larger data area.

– Double sided discs, actually two platters bonded back to back, increasing rigidity and minimizing laser focus problems due to warpage. This doubles the disc's storage capacity.

– Two layers of data per platter, where the laser changes focus to access the different layers.

– More efficient error correction and video compression algorithms.

Not all DVDs hold 17 gigabytes of data because they aren't configured the same. Different configurations include:

– DVD 5   single sided, single layer   4.7 gigabytes

– DVD 9   single sided, double layer   8.5 GB

– DVD 10   double sided, single layer   9.4 GB

– DVD 18   double sided, double layer   17 GB

All DVD players have capacity to read all the above formats. Some players require the user to remove the disc and turn it over, because the manufacturers are too cheap to install separate lasers for both sides.

Many of today's larger sessions take up so much space, DVDs are used to store the data. Gone are the good old days when a simple CD would hold all your data.

• **What is the difference in actual sound between digital recording and analog recording?** It is not unlike the difference between watching a film (analog) and watching a video (digital). You can see the difference, but try explaining it to someone. Differences include:

– With analog, some lower frequencies may be slightly accentuated, adding a 'warmth' not captured with digital.

– Peak levels can oversaturate an analog tape. Some engineers use this as part of their sound.

– A slight amount of high end may be lost on analog playback, simply due to limitations of analog tape. Digital recording processes audio with no perceivable loss.

– With digital recording, there is no signal loss due to tape shedding, no crosstalk, no track leakage, and no such thing as outside track degradation, commonly found on tracks 1 and 24 on analog.

– Digital non-audio tracks may be available for synchronization. An 8 track ADAT machine will have 8 audio tracks, plus at least one more track used for synchronization. 8 track analog is just that – 8 tracks. Use them how you will.

– Analog copying will always have generation loss. There is no loss in digital because a copy is a clone of the ones and zeros.

– Convenience. Today, with removeable hard drives, a complete session can easily be taken from studio to studio. And no need to download. You can just work off the hard drive.

– Traditionally, the engineer relied solely on his ears. Now he can see the waveforms and hear the sound to help him place it in the right spot.

Index

Assistant Engineers Handbook
Tim Crich

- ❏ Over 275 pages
- ❏ Fully illustrated
- ❏ Packed with proven recording studio secrets
- ❏ Complete setups for basic, overdubs and mixing
- ❏ Proper microphone handling and setup
- ❏ Most efficient ways to complete all paperwork
- ❏ Needed pointers on recording studio etiquette
- ❏ Established procedures to keep the session moving
- ❏ Key priorities for before, during and after the session
- ❏ Essential tips on setup and breakdown of all equipment
- ❏ Full chapter on tape machines and alignment
- ❏ Required reading in audio schools throughout America
- ❏ Much, much more!

About the Author

Tim Crich began his career as an illustrator in Western Canada, but moved to New York to pursue a career in recording. With over twenty years of experience in the recording studio, Tim has credits on some of the biggest records in history, and has engineered for the biggest producers in the world. Tim is a recording engineer and writer/illustrator living in Vancouver, Canada.

For further details and to order visit **www.aehandbook.com**

ISBN 0-969-82230-8

 **Focal Press**   **www.focalpress.com**

Join Focal Press online
As a member you will enjoy the following benefits:

- browse our full list of books available
- view sample chapters
- order securely online

Focal eNews
Register for eNews, the regular email service from Focal Press, to receive:

- advance news of our latest publications
- exclusive articles written by our authors
- related event information
- free sample chapters
- information about special offers

Go to www.focalpress.com to register and the eNews bulletin will soon be arriving on your desktop!

If you require any further information about the eNews or www.focalpress.com please contact:

USA
Tricia Geswell
Email: t.geswell@elsevier.com
Tel: +1 781 313 4739

Europe and rest of world
Lucy Lomas-Walker
Email: l.lomas@elsevier.com
Tel: +44 (0) 1865 314438

Catalogue
For information on all Focal Press titles, our full catalogue is available online at www.focalpress.com, alternatively you can contact us for a free printed version:

USA
Email: c.degon@elsevier.com
Tel: +1 781 313 4721

Europe and rest of world
Email: j.blackford@elsevier.com
Tel: +44 (0) 1865 314220

Potential authors
If you have an idea for a book, please get in touch:

USA
editors@focalpress.com

Europe and rest of world
ge.kennedy@elsevier.com